Environmental Issues for the Gulf

Environmental Issues for the Gulf

Oil, Water and Sustainable Development

Edited by Peter Kassler

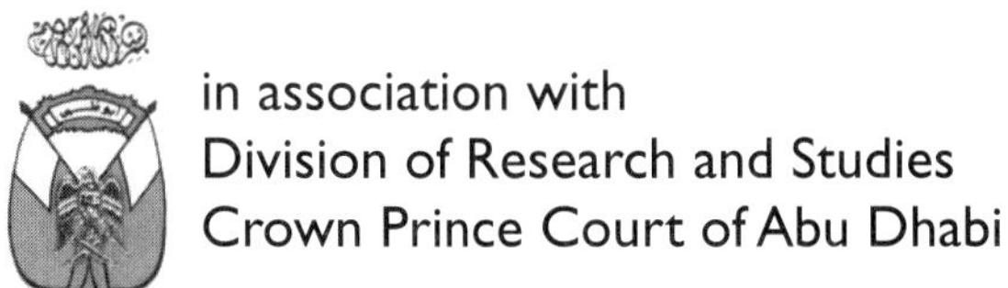

First published in Great Britain in 1999 by
Royal Institute of International Affairs, 10 St James's Square, London
SW1Y 4LE
(Charity Registration No. 208 223)

Distributed worldwide by
The Brookings Institution, 1775 Massachusetts Avenue NW,
Washington DC 20036-2188, USA

British Library Cataloguing in Publication Data
A CIP catalogue record for this book is available from the British Library.

Hardback: ISBN 1 86203 029 4
Paperback: ISBN 1 86203 024 3

Typeset in Times by Koinonia
Printed and bound in Great Britain by Alden Press
Cover design by Shailesh Chavda

Contents

Contributors

Dr Saif M. Al-Ghais is Secretary General of the Environmental Research and Wildlife Development Agency in Abu Dhabi. He acquired a degree in Biological Sciences from Seattle Pacific University, USA and a doctorate in Marine Biology from the University of Liverpool, UK and thereafter has been associated with the teaching and research programmes of the UAE University in biological and environmental sciences as a teaching staff member of the Biology Department and the Head of the Marine Environment Research Section.

Dr Al-Ghais is a marine biologist dedicated to the cause of environmental conservation and sustainable management of natural resources in the UAE. He has published widely and is well known for his contribution to pioneer projects such as the characterization and compilation of fish species inhabiting the UAE waters, the rehabilitation and conservation of the mangrove ecosystem, the conservation of sea turtles and the impact assessment of environmental pollution. He was awarded research grants by Shell International in the UAE to support mangrove and sea turtle conservation projects.

Roger Cragg is a principal engineer in the water design group of Hyder Consulting. He is a chartered engineer and a member of the Institution of Civil Engineers and of the Chartered Institution of Water and Environmental Management. He has worked exclusively on water projects for over 26 years, gaining extensive experience of both water supply and sewage treatment projects throughout the Middle East. His work in Qatar, Abu Dhabi, Oman, Saudi Arabia, Yemen and Bahrain has provided him with detailed insight into issues relating to water resources, water use and effluent disposal as they affect the Middle East.

Dr Michael Grubb is an Associate Fellow with the Energy and Environment Programme at the Royal Institute of International Affairs (RIIA) and Visiting Professor at Imperial College, London. He was Head of the Energy and Environmental Programme at the RIIA from January 1993 to

August 1998, leading the Programme to become one of the world's premier centres of research on international energy and environmental policy issues.

Dr Grubb is well known for his work on the policy implications of climate change and on renewable energy sources. He was a Lead Author for the Second Assessment Report of the Intergovernmental Panel on Climate Change, convening the chapter on equity and social considerations, and has advised many international study groups and organizations. He is currently a Convening Lead Author for the IPCC's Special Report on Technology Transfer and a lead author on the international UNCTAD study on guidelines, rules and modalities for greenhouse gas emissions trading. He is a member of the Advisory Board to the President of the International Association for Energy Economics, the editorial board of *Energy Policy*, and the UK Government's Green Globe Task Force.

Dr Grubb's publications include a 1989 report on the options for international climate change negotiations that helped to launch a global debate on the use of tradable emissions quotas for climate control; a two-volume international study entitled *Energy Policies and the Greenhouse Effect*; and a range of journal publications on economic and political aspects of the problem. He has also led an extensive study resulting in a book on *Emerging Energy Technologies*, and published a book examining the outcome and implications of the Rio 'Earth Summit'. His most recent (co-authored) book, *The Kyoto Protocol: A Guide and Assessment*, was published in June 1999 and is being translated into Japanese, Russian and Farsi.

Dr Peter Kassler is a Consultant and Associate Fellow with the Energy and Environmental Programme at the Royal Institute of International Affairs. Dr Kassler left Imperial College, University of London, in 1960 with a PhD in geology. From 1960 to 1995 he had an international career with Shell, initially as an earth scientist in oil and gas exploration and production in the Middle East and elsewhere, then in strategic planning and general management positions. His last two jobs were as Middle East Coordinator and Group Planning Coordinator. In the former he was responsible for all Shell's business activities in the Middle East, South Asia and French-speaking Africa. In the latter he managed a small division responsible for strategic advice to the Chairman of the Royal-

Dutch/Shell Group and all the Shell businesses and regions. The planning division included the Shell Scenario Group, well known for developing some unusual and challenging visions of the long-term future.

Since retiring from Shell at the end of 1995 Dr Kassler has worked as an independent consultant, one of his projects being a study at the RIIA on the impact of climate change policies on energy-exporting countries. The results of this work were published in *Energy Exporters and Climate Change* in 1997. He is also active as a business and strategy consultant in the oil and gas industry. Since mid-1997 he has been working with British-Borneo Petroleum Syndicate PLC, an independent British oil company with exploration and production activities in the deep-water Gulf of Mexico, the North Sea and elsewhere. His role is as a senior strategic adviser to the Chief Executive, particularly on exploration matters.

Brent Pyburn is currently Manager of Emergency Response for BP Shipping. He is also Vice-Chairman of the International Petroleum Industry Environmental Conservation Association Oil Spill Working Group and Chairman of the Global Initiative subgroup. After a successful seagoing career with BP Shipping and attaining his Master's Foreign Going Certificate he spent two years in the Shetland Islands as berthing Master at the then new Voe Terminal.

In 1981 he had a major role in the establishment of the world-famous Oil Spill Service Centre in Southampton, managing it from 1985 to 1989. In 1989 he led a team of experts to Alaska to assist Exxon during the *Exxon Valdez* incident and subsequently on his return to the UK developed a Crisis Management Organization for the BP Group worldwide, training the BP Board of Directors and senior executives. In 1992 he was asked by five oil majors to establish and manage a regional oil spill response centre, East Asia Response Ltd (EARL), based in Singapore. He was Chief Executive of East Asia Response Ltd until returning to the UK in 1997. Mr Pyburn was awarded an MBE in 1990 for his services to oil spill response and technology. He was also made a Fellow of the Institute of Petroleum in the same year.

Roger Rainbow recently retired from Shell where he was Vice-President, Global Business Environment, in Shell International Ltd. He studied at Oxford University and the London Business School and joined Shell in

1970. During his career with Shell he worked in the Philippines, Brunei, Turkey, Venezuela, Australia and London. The responsibilities of his last team were to provide analysis of the business environment, including the preparation of long-term scenarios. Topics covered included the economic outlook, long-term energy supply and demand, and social, technological and environmental issues.

Since his retirement, he has continued to consult on scenarios and business strategy. He is Strategic Planning Adviser for the Control Risks Group, an Associate of the Center for Generative Leadership, a US strategy and change consultancy, and is a member of the Directorate of Studies of the Windsor Leadership Trust, a charity that helps develop the leaders of tomorrow's Britain.

Hywel Thomas is Director of Water Processes for Hyder Infrastructure Developments. In this role he is the company's technical leader for the water, wastewater and related environmental management consulting business. On graduating in civil engineering in 1960, he immediately entered the water industry with consulting engineers in London. Progression through various organizations in the public sector led to a return to private-sector consulting work with the privatization of the UK water utilities in 1989. Subsequent acquisitions by the company have meant changes in the company name, finally resulting in the recent rebranding to Hyder. That part of the company with extensive experience of water-related business in the Gulf was previously known as Acer John Taylor.

List of seminar participants

H.H. Sheikh Saeed Bin Saif Al Nahayan, Director, Research and Studies Division, Crown Prince Court

Dr Adham Mahdi Abdullah, Economic Adviser, Public Industrial Establishment

Dr Mohammed Abdul Rahman Al Assoumy, Director, Emirates Industrial Bank

Dr Saif Mohammed Al-Ghais, Secretary General, ERWDA

H.E. Abulmalik Al Hamar, Adviser, Presidential Court, UAE

Ahmed Al Harmoudy, Board Member, Environment Friends Society

Dr Sulaiman Al Jassim, Director, Higher Colleges of Technology

Hassan Saeed Al Kuthairy, Deputy General Manager, Food and Environment Control Center

Abdullah Usama Al Maleki, Economic Adviser, UAE Central Bank

Majid Al Mansouri, Environment Protection Adviser, ADNOC

Mohammad Saeed Al Muhairy, Board Member, Environment Friends Society

H.E. Ahmed Ali Al Sayegh, Board Member, ERWDA

Adel Saeed Al Shamsi, Emirates Affairs Head, Crown Prince Court

Dr Ahmed Abdullah Al Tamimy, Industrial Section, Planning Department of Abu Dhabi

Yusri Al Tamimy, Adviser of Corporate Planning, ADNOC

Dr Salwa Al Tayeb Al Akraa, Society Member, Environment Friends Society

Dr Iskandar Bashir, Adviser, Crown Prince Court

H.E. Jasim Darwish, Secretary General of Municipalities

Dr Mohammad Diffrawi, Representative, Shell Gas Company

Dr Michael Grubb, Associate Fellow, Energy and Environment Programme, RIIA

Dr Sabri Hasanein, Economics Expert, Crown Prince Court

David G. Heard OBE, Representative, Abu Dhabi Petroleum Co. Ltd

Dr Frauke Heard-Bey, Centre for Documentation and Research

Dr Rosemary Hollis, Head, Middle East Programme, RIIA

Dr Saeed Mushtak Hussein, Chief Technical Adviser for Planning Department, Planning Department of Abu Dhabi

Dr Peter Kassler, Consultant and Associate Fellow, Energy and Environmental Programme, RIIA

H.E Adel M. Khalifa, UN Resident Coordinator, Abu Dhabi

Bashir Ahmed Khudur, Manager, ERWDA

Munif Othman, Senior Investment Manager, ADIA

Brent Pyburn, Manager of Emergency Response, BP Shipping

Roger Rainbow, Vice-President, Global Business Environment, Shell International *(retired)*

Dr Abdullah Abu Ruwaida, Public Health Adviser, Municipalities General Secretariat

Dr Mohammed Shehab, UNDP Strategic Study Coordinator, Planning Department of Abu Dhabi

Jack Shiller, Health Affairs, ADNOC

Martin Smize, Environment Protection Adviser, ADNOC

Abdulkareem Thabet, Marketing Department, ADNOC

Hywel Thomas, Director of Water Processes, Hyder Infrastructure Developments

Dr Naif Ubaid, Political Expert, Crown Prince Court

Jibril Mohammed Zeyadah, Legal Adviser, Crown Prince Court

Abbreviations

ADCO	Abu Dhabi Company for Onshore Oil Operations
ADIA	Abu Dhabi Investment Authority
ADNOC	Abu Dhabi National Oil Company
AOSIS	Association of Small Island States
CCGT	combined-cycle gas turbine
CDI	Chemical Distribution Institute database
CLC	Civil Liabilities Convention
CO_2	carbon dioxide
ERWDA	Environmental Research and Wildlife Development Agency (Abu Dhabi)
FEA	Federal Environment Agency (UAE)
GATT	General Agreement on Tariffs and Trade
GCC	Gulf Cooperation Council
GDP	gross domestic product
GHG	greenhouse gas
GWP	global warming potential
IEA	International Energy Agency
IMO	International Maritime Organization
IPIECA	International Petroleum Industry's Environment Conservation Association
IPCC	International Panel on Climate Change
MARPOL	International Convention on the Prevention of Pollution by Ships
NARC	National Avian Research Centre (Abu Dhabi)
NEBA	net environmental benefit analysis
NGO	non-governmental organization
OAPEC	Organization of Arab Petroleum Exporting Countries
OCIMF	Oil Companies' International Marine Forum
OECD	Organization for Economic Cooperation and Development
OPEC	Organization of Petroleum Exporting Countries
OPRC 90	International Convention on Oil Pollution Preparedness Response and Cooperation 1990

PNGV	Programme for the Next Generation Vehicle
R&D	Research and Development
SIRE	Ship Inspection Report
UAE	United Arab Emirates
UNCTAD	United Nations Conference on Trade and Development
UNDP	United Nations Development Programme
UNEP	United Nations Environment Programme
UNFCCC	United Nations Framework Convention on Climate Change
VLCCs	very large crude carriers
WBCSD	World Business Council for Sustainable Development
WEC	World Energy Council
WMO	World Meteorological Organization

Foreword

This work is the product of a collaborative endeavour in several respects. It is a joint undertaking between two research institutions, with contrasting national and cultural perspectives, and brings together the insights of experts from very different disciplines, with career experiences ranging from seafaring to civil engineering to corporate planning. Perhaps most remarkable, here is a discussion of some of the thorniest issues to face Gulf energy producers, and the initiative to explore all of them came directly from the region, from Abu Dhabi.

The book is the second to result from cooperation between the Middle East Programme at the Royal Institute of International Affairs and the Research and Studies Division of the Crown Prince Court in Abu Dhabi. The first tackled the subject of oil and regional developments in the Gulf, and was published in 1998. In both cases the edited volumes are based on papers originally given at seminars in Abu Dhabi, and revised in light of the discussions which took place with the relevant experts and interested parties from the United Arab Emirates (UAE). The topics addressed in the first instance had to do with the implications of regional and international political developments for the energy scene. The contributors discerned a mixture of factors at work, including socio-economic trends in Iran, Iraq and Saudi Arabia, and policy choices of the United States and Europe, which have a bearing on how much oil and gas is produced and by which countries. Here the focus is on environmental concerns which affect Gulf energy producers in general and the UAE in particular.

The result is a reference source on the latest thinking about sustainable development, global warming, water and waste management, as well as methods and agreements for responding to climate change and oil spills. The UAE already has a programme for environmental protection, on land and offshore, as described by one of the contributors here. However, as is also made clear, that does not mean that everything is taken care of, since only some of the problems are local and amenable to direct action, while others are of a regional dimension and can only be tackled in cooperation with neighbours. Yet other concerns are global, and beyond the control of any single country, though each is now expected to have a position on the question of climate change and the various remedies proposed.

A number of specialist issues, such as emissions trading and desalination technology, are touched on in this book, but the treatment is accessible and helpful to the lay reader. In fact, this work will be of as much value to those interested in Middle East politics as to those whose preoccupations are energy and the environment. Meanwhile, the discussions about sustainable development are an adventure in lateral thinking. Speaking as a political scientist, I found the whole project both stimulating and useful.

On that note, in my capacity as Head of the Middle East Programme at the RIIA, I should like to thank our partners in this project, the Research and Studies Division of the Crown Prince Court in Abu Dhabi, for making it possible, and in particular H.H. Sheikh Saeed Bin Saif Al Nahayan, for identifying the focus of the work and for his attention and support during the process. Dr Iskandar Bashir deserves special thanks for coordinating and facilitating our joint endeavours with the Crown Prince Court. At Chatham House, meanwhile, two people have been crucial to the publication process – Raksha Thakor, administrator of the Middle East Programme and Margaret May, Head of the Publications Department.

May I also thank all those who contributed to the seminar and this volume, and especially Peter Kassler for accepting the task of editor and drawing the threads and ideas together. The views expressed in each of the papers here are those of the authors alone, and responsibility for summarizing the main points to come out of the discussion at the seminar lies with the editor.

August 1999

Dr Rosemary Hollis
Head, Middle East Programme
Royal Institute of International Affairs

Map of the region

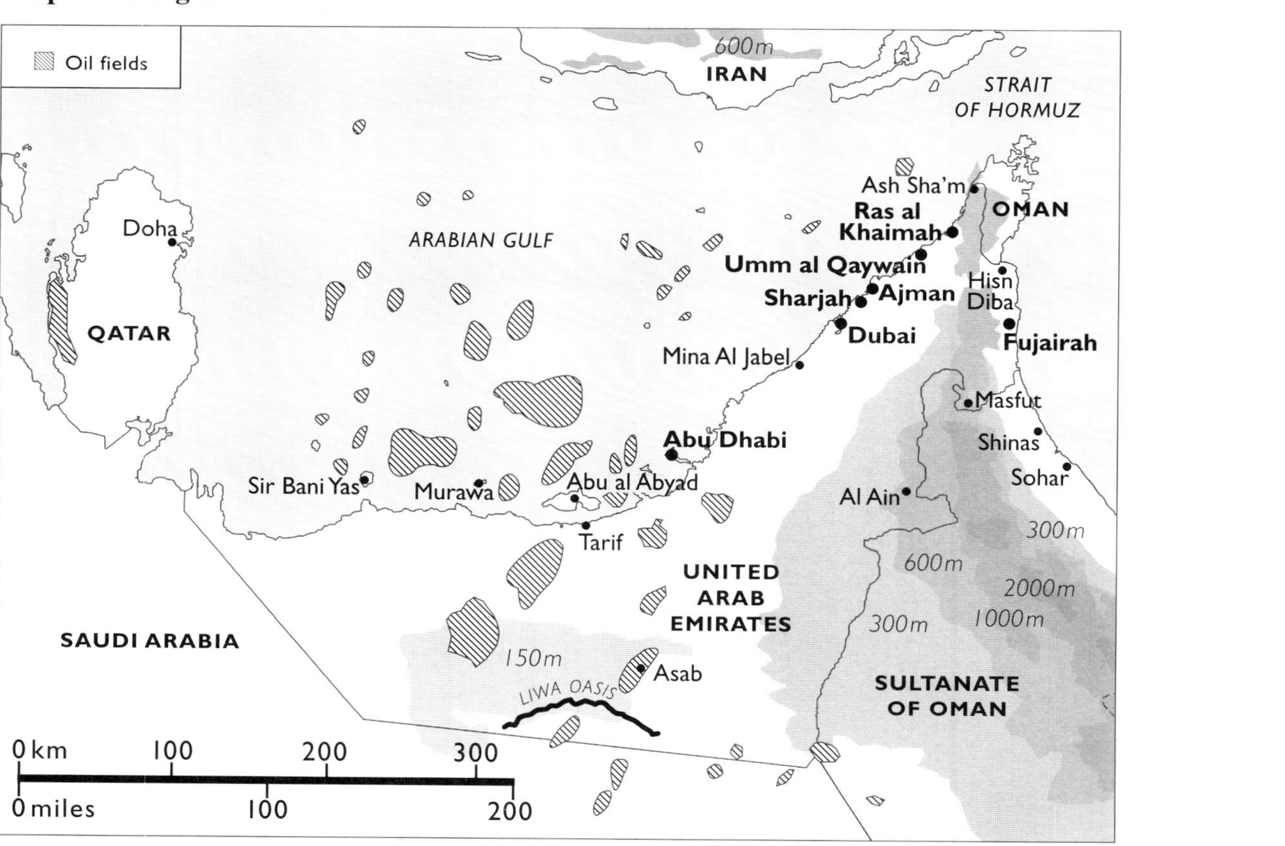

1 Overview

Peter Kassler

This book is about sustainable development, an idea which has been given various definitions; for our purposes we decided to work with that of the Brundtland Commission, set down in 1987 in its report *Our Common Future*: development that 'meets the needs of the present without compromising the ability of future generations to meet their own needs.'

The papers included in this volume are based on presentations and discussions which took place at a seminar in Abu Dhabi in October 1998, which was jointly undertaken by the Abu Dhabi Crown Prince Court and the Middle East Programme of the Royal Institute of International Affairs. The context of this work on sustainable development ranges from the global to the local, but the object was to illuminate issues of interest to the Emirates.

It is only 27 years since the United Arab Emirates became a new nation-state under the leadership of His Highness Sheikh Zayed Bin Sultan Al Nahayan, the current president. In that sense only, this is a very young country; but in reality it has a long and colourful history, largely recorded in the oral traditions of its people and those who settled there. The period since 1972 has been one of almost unimaginably rapid change, brought about by the rise of the oil industry and its economic consequences for the Emirates, particularly for Abu Dhabi. Oil and gas have brought great wealth and prosperity to the country. Moreover, although the old Trucial States were noted as centres of regional trade and were by no means entirely inward-looking, the arrival of oil has brought the Emirates out on to the global scene as a country with world-class investors and trading partners, and as a name to be reckoned with both in the oil and gas industries and in regional and international politics. During this process, UAE residents have been exposed to all the influences of the Western media and to models of consumer society which were simply not available in the days before oil. Nevertheless, and despite all these influences, the UAE is widely respected for the quality of the development path which it has elected to follow and for its leaders' skill in finding accommodations between the traditional and the new.

This raises one of the most important dimensions of sustainability for the Emirates: the preservation and renewal of the historical cultural and ethical values of the region, in the face of new industrial and economic developments and the arrival of large populations of immigrants and expatriates with different backgrounds, including other Islamic traditions. This particular aspect of sustainability was perhaps not directly addressed by the papers presented at the seminar which led to the writing of this book. It nevertheless underlies much of the discussion that took place and has always been a prominent factor in the determination of public policy in the UAE.

Another set of factors to be balanced relates to the development of the UAE's oil and gas resources, the majority of which are located in Abu Dhabi. Here the notion of sustainable development has a particular resonance, since these are non-renewable resources. This source of wealth, only a quarter of a century old, is being used to fund the development of a modern economy and its social institutions. Every barrel of oil, every cubic foot of gas, can be produced and consumed just once; then it is gone. This means that almost every important decision to do with the management of the oil industry or the running of the economy contains an element of intergenerational judgment. For example:

- the timing of oil and gas production (the price may go up or down; new technologies or changes in consumers' preferences may greatly increase or decrease the value of the resource for the next generation);
- diversification of the economy (it may be dangerous in the very long run to become over-dependent on one group of commodities, but most of the feasible alternatives to oil yield lower returns on investment, and the conditions for attaining commercial advantage in new industries are not necessarily present);
- investment in education (How much? Where? In what skills to provide for the next generation's society and economy? In what numbers to supply the next generation's employment needs?);
- management of food and water supplies (encouraging local agriculture, providing fresh food and economic support to farming

communities but consuming scarce water, the alternative being to import food from other countries in the region).

These examples all depict trade-offs between present needs and future welfare, or between resource demands aimed to satisfy different sustainability goals. In this sense there is no free lunch. In a static world, most decisions to dedicate more resources to one activity or purpose also imply doing less of something else. This is true as long as available resources remain finite; however, the one factor which can increase resources is economic development.

Turning to climate change, we see a further example of the forces pulling the Emirates in opposing directions. The oil industry has been targeted by environmentalists as the villain of the piece in delivering large quantities of greenhouse gas-generating fuels to the world energy market (although papers in this volume show convincingly that coal is much more of a problem than oil in this respect). However, it is also clear that the revenues from oil are and will continue to be essential in helping the Emirates to attain their own version of sustainability. Reducing these revenues in the interests of the global environment will create local problems in energy-exporting countries which will need to be managed.

The ideas expressed in Roger Rainbow's paper (Chapter 2) about the differing requirements of sustainability at different points in the development cycle are helpful in finding a way through these dilemmas. One conclusion is that most developing countries need to develop their own local economies and societies and manage their local environmental problems before the global issues start to have a legitimate claim on their attention. Development can therefore be made to lead to sustainability, even if the short-term goals of sustainable development differ between, say, Sweden and the Emirates.

In any case, whatever changes result from climate change policy, the world will go on needing the UAE's oil and gas for a number of years. For how long, exactly, is uncertain and depends on global developments, not local ones. The Emirates have the advantage of possessing reserves of natural gas and perhaps prospects of discovering further resources, and (as advocated in two of these papers) gas is likely to be an extremely valuable low-carbon fuel of transition to ease the long passage from today's fossil-fuel-based economy to whatever succeeds it. However, today is not too soon to start thinking about how the ultimate decline of oil will affect the Emirates and their neighbours, and what are the best actions to safeguard their economic health, social institutions and environments for future generations.

The papers delivered at the seminar in October 1998 were designed to illuminate some aspects of these choices. As a result of the growth of the past quarter of a century, the Emirates are now faced with questions about sustainability at three more or less distinct levels – global, regional and local – although several of the papers range across more than one level.

The global sustainability issues with which the Emirates have to deal are mainly to do with the UAE's important position in the world oil market, its dependence on oil as a source of revenue and the possible impacts of the changes referred to above on its welfare. Papers in this book by Roger Rainbow, Michael Grubb and Peter Kassler (Chapters 2–4) cover different aspects of this question. It is difficult or impossible for the UAE to stand alone on such issues and it has to act as a member of international organizations, including OPEC, OAPEC, and the GCC. These groups of countries, with shared and clearly defined interests and negotiating positions on important single issues, are capable of having an impact in international negotiations.

A number of issues affecting the UAE have a regional span (referring approximately to the Middle East). These include water supply, and the interrelated topics of marine pollution and fisheries. Water-related problems are probably partly local and partly regional, although they might have a global impact if political disturbances resulted from food and water shortages. Roger Cragg and Hywel Thomas makes it clear in their paper (Chapter 6) that the UAE's indigenous fresh water resources are not sufficient to sustain the current population and its present activities. Possible solutions include more efficient use of existing supplies. Beyond this option, new supplies could come from further desalination, perhaps using renewable energy sources, or in effect by importing water embodied in food from other countries in the region. A separate water-related issue, and one which poses a significant threat, is the impact of abstraction of waters from the rivers Tigris and Euphrates, by Turkey and other countries, on the water circulation in the Gulf and its natural habitats.

In Chapter 7 Brent Pyburn describes a responsible oil-company approach to preventing oil spills, but is realistic in taking the view that it is very difficult to legislate against human error and that the prevention system therefore needs to be backed up by effective minimization of the results of spills that do occur. Small local spills can clearly be managed at the loading terminal or other installation where they occur, but in a closed sea like the Gulf there must be careful contingency planning among the littoral states to deal with incidents affecting more than one country's waters. Since the tanker fleets move through the entire Gulf, any effective monitoring of polluting vessels needs to be coordinated at a regional level.

Commercial fish stocks and fishing vessels also move easily across international boundaries, and management of offtake levels cannot effectively be achieved by one country alone. This issue, discussed by Saif Al-Ghais in Chapter 5, is therefore another that calls for a regional approach.

At the local level the issues of sustainability are to do with the conservation of the local environment and the management of resources to be used by Emirates residents. These are covered by Saif Al-Ghais, who eloquently describes the problems confronting the terrestial and marine floras and faunas and their habitats in the face of encroachment by industry and expanding human residential areas. These local environmental questions are the ones which fall most clearly within the remit of the Emirates authorities and their institutions; they are also ones where the formation of public opinion and the involvement of residents can be most important. Management and improvement of ecological strategy were taken up by the president and the Emirates municipalities at an early stage in the existence of the UAE, exemplified by the plantations of bushes and trees adorning the streets of the main cities, as well as by the setting up of offshore wildlife conservation areas.

The six papers in this volume were commissioned by the Crown Prince Court and the Royal Institute of International Affairs to illuminate different aspects of sustainable development affecting the Emirates. There are probably other topics which could have been covered, but these are believed to be the most important ones. It is obvious that the UAE has gone through tremendous development since 1972, certainly economically but also socially, in its attitude to the environment and its standing in the world. These changes have been facilitated by access to great wealth, but the UAE is also a striking illustration of a point made by Roger Rainbow: the mismatch in most societies between the typical four- to seven-year terms of office of their leaders and the much longer lifetimes of major projects, whether of infrastructure construction, social reform or environmental conservation. The UAE has been fortunate enough to have had stable leadership through its entire existence to date and the results are evident. Because of the country's oil reserve base and low production cost, it is probably one of the most robust oil exporters and has more time to plan its future than most others. Even so, the message of the seminar and these papers is that the time to start is now.

2 Sustainable development: a view from an oil major

Roger Rainbow

As the world has become richer and its population has increased, it is clear that a single indicator such as GDP per capita is not a sufficient indicator of well-being. Growing concern that economic development may happen at the expense of other considerations, particularly the health of the environment, has led to the development of the concept of sustainable development.

This paper looks at the concept from the point of view of a major oil company. It is hoped that this perspective will also be helpful for other actors, in particular countries with a high dependence on oil and gas.

The definition of 'sustainable development' adopted for the purposes of this book, that published by the Brundtland Commission in 1987, is perhaps the clearest. In defining the term to mean development that 'meets the needs of the present without compromising the ability of future generations to meet their own needs', it immediately introduces one of the key underlying ideas, that of intergenerational equity. The effects of what we do today may be felt only years, decades or even longer in the future, and these effects may be profound. The challenge is that most of our institutions, in particular economic and political decision-making institutions, tend to work on much shorter time horizons.

The three domains

There are three broad areas, or domains, that underlie the analysis of sustainable development: economic development, the state of the environment, and the development of society. We will briefly look at some of the issues relating to each domain in turn; but – and this is a point that goes right to the core of the debate – it is the linkages and interactions among these domains that are most important. We have to look at things holistically.

Myths and mental maps

At this stage it is worth going a bit further. The World Business Council for Sustainable Development is an organization of business firms from around the world, committed to the concept of sustainable development, which have got together to help one another understand the issues and begin to implement action. They have recently published some scenarios on sustainability. One of their insights (Figure 2.1) is that the way different individuals or organizations see an issue as complex and as potentially divisive as sustainable development depends a lot on their own view of the world – what scenario planners call a 'mental map'. The WBCSD team used the concept of myths.

Some people see the world and all its problems as subject to and soluble by science and analysis. Reasonable people can sit down together and come to the truth. This is the scientific myth. Others see things very

Figure 2.1: Sustainability domains

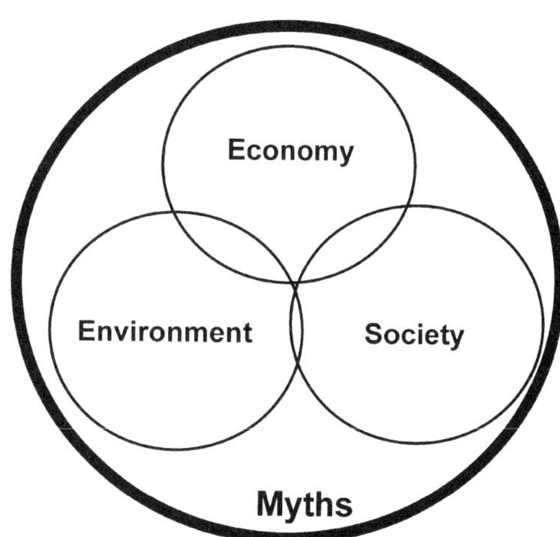

Source: WBCSD Sustainable Development Scenarios, 1997.

much in economic terms. Virtually all human transactions can be described and predicted by discovering the incentives that lie behind them. At its widest this myth sees people as 'coin-operated machines'. For yet others, their view of the world and its affairs is governed by a higher imperative not subject to humdrum analysis. This religious myth obviously covers the teaching of the great world religions, such as Islam or Christianity, but also extends, for example, into debates on the environment. Many environmentalists see planet earth as itself an almost living organism, complex and self-organizing. This concept of Gaia can lead to a view that any damage to the environment is indefensible. Finally, there is the hero myth. Some people interpret events and developments by focusing on the players, the heroes and villains. The success of companies is related to the success of the individual CEO. Geopolitical issues are described in the language of 'the evil Saddam' or 'the Great Satan of the US'.

All of these myths have validity, and it is essential to understand them to understand the development of the debates over sustainable development. Moreover, if companies or other players wish to have an effect on the overall debate, it is essential they understand the myths of other participants and address their concerns.

As an illustration, we can use the episode of the Brent Spar offshore facility. Shell UK did a detailed and painstaking analysis of the costs and benefits of various ways of disposing of the Brent Spar. After consulting with many experts, including marine biologists, they came to the conclusion that the best solution was to sink the platform in the deep Atlantic. In our myth terms, this was the scientific myth at its best. But when Greenpeace occupied the Brent Spar, they only half-heartedly attacked the scientific analysis. The appeal they made, particularly in Germany, the Netherlands and Scandinavia, was to a religious myth. Sinking the platform was a desecration of nature and a sin. It was not a question of optimization and balance, but of good and evil. And then there was the media reaction – without which nothing much would have happened. The media, particularly television, tend to see events in terms of heroes and villains; and for them, this was a typical David and Goliath story. Plucky young activists were taking on the mighty, multinational and selfish oil company. The heroes and villains were clear. In this context, restating the scientific case was useless; it merely sounded pompous and uncaring.

Understanding the myths and mental maps of various players is as important as understanding the issues of the domains and their linkages.

The economic domain

This is perhaps the best-known of the domains and the easiest to consider. We know more or less how to measure economic activity, by GDP or consumption of goods and services. It is important to understand that economic development is central to sustainable development. Sustainable development is not a paper cloak for those who are against the world getting more prosperous.

Economic development is a relatively recent concept. It is really only in the last 250 years or so that people have come to accept that increasing prosperity per capita is possible; today it is widely regarded as a right. Economic growth was given huge impetus by the industrial revolution and it spread globally; the phenomenon of worldwide growth is more recent still, having markedly accelerated since the end of the Second World War. It has far outstripped population growth. Growth in the world economy is illustrated in Figure 2.2.

Figure 2.2: World economic growth, 1500–1990

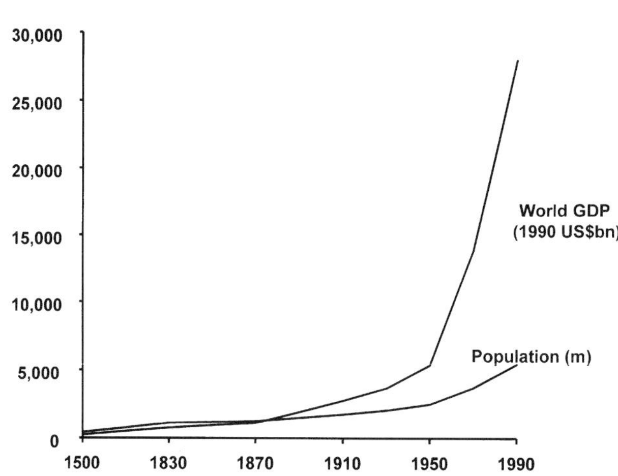

Source: Angus Maddison, *Monitoring the World Economy 1820–1992* (Paris: OECD, 1995)

Energy, of course, has been one of the key drivers of this surge of growth. The industrial revolution would not have been possible without coal and steam, and today's level of economic development would not have been possible without oil and gas.

This quest for prosperity has led to the development of several new forms of institution. In particular, the limited liability, joint stock company was a development of the nineteenth century. Corporations such as Shell owe their form to a deliberate, and successful, invention of a legal framework by society to allow them to be principal actors in creating wealth. Alternative models such as communism that did not allow such private companies have proved unsuccessful in meeting the economic (and indeed many other) aspirations of their peoples.

Of course, this overall success masks continuing problems. Perhaps most important are the inequalities in performance and outcome. Figure 2.3 shows two examples. One highlights the disparity between the rich world of the OECD and a relatively poorly performing region such as the Middle East and North Africa. For a while, such inequalities were seen as a manifestation of great injustices in the world system, including imperialism and its supposed child neo-colonialism. But the other graph in this figure tells a different story. Poor countries can get richer, and inequality of performance has a lot to do with domestic economic policy and social development, including education.

Figure 2.3: Uneven distribution of income growth, 1965–95

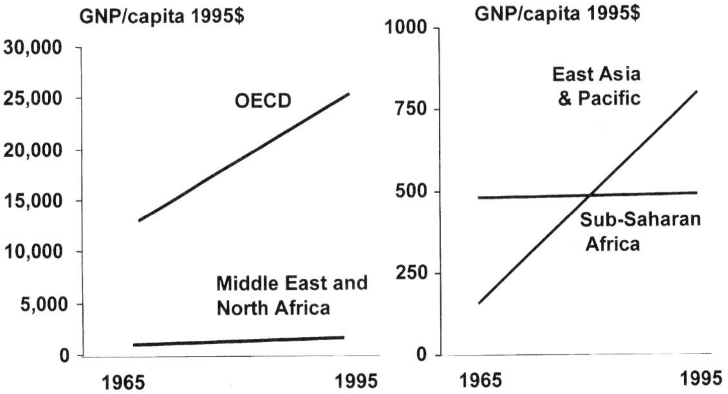

Source: Data from IBRD, 1997.

This inequality also manifests itself in the demand for energy. One-fifth of the world's population uses almost two-thirds of its electricity.

The reason for focusing on inequality as an issue is that it is one of the key factors that may determine whether a model of economic development is sustainable. Clearly, within a society, it is not always pleasant to be rich when surrounded by a large mass of angry poor. And needless to say, it is of little comfort to be very poor and starving and be told your country is rich.

As we shall discuss later, the other key issue in the economic domain is the link between economic development and the destruction of the environment.

The environment

It is a widely held view that as the world has got richer, so it has got more polluted. Certainly, as Figure 2.4 shows, with increases in the world's population and income have come greater pressure on natural resources, both renewable and non-renewable. These include water, fish and forests.

Figure 2.4: State of the world environment, 1950–97

	1950	1972	1997
Population billion persons	2.5	3.8	5.8
Megacities cities with population greater than 8 million	2	9	25
Food average daily food production in calories/capita	1,980	2,450	2,770
Fisheries annual fish catch in million tons	19	58	91
Water Use annual water use in cubic kilometres	1,300	2,600	4,200
Rainforest Cover index of forest cover, 1950=100	100	85	70
Elephants million animals	6.0	2.0	0.6
CO_2 Emissions annual CO_2 emissions in billion tons of carbon	1.6	4.9	7.0
Ozone Layer atmospheric concentration of CFCs in parts/billion	-	1.4	3.0

Source: World Resources Institute, *World Resources 1996–97: A Guide to the Global Environment* (Oxford University Press, 1996).

Also, there has been a reduction in pristine areas of the world, a reduction in many animal species and a fear of an accelerating loss of species. This is particularly worrying for those who see this issue through the eyes of the religious myth. Man is seen as the plunderer and destroyer, in complete opposition to the eighteenth- and nineteenth-century view that man's triumph was in the mastery of nature.

In terms of pollution, the interactions between development and the environment are more subtle. The World Bank in a 1992 report broke pollution issues down into three broad categories (Figure 2.5): problems of the poor, problems of development and problems of affluence.

Figure 2.5: Development and the environment

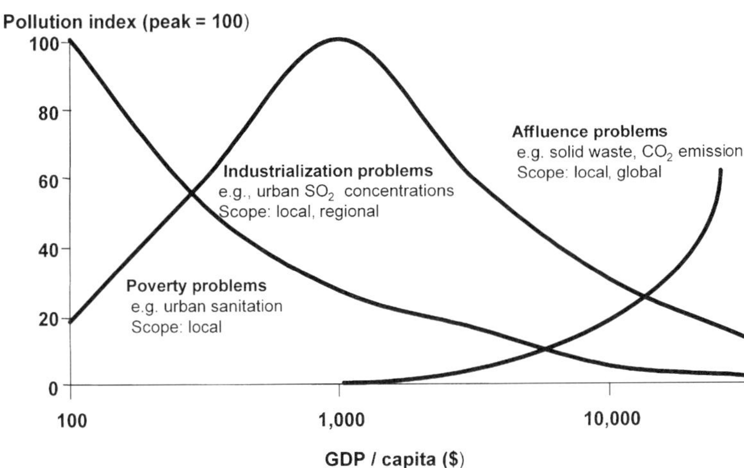

Source: World Bank, *World Development Report*, 1992.

Problems of the poor Pollution problems of the poor, such as dirty water and lack of sanitation, are deadly. Perhaps 100 million people a year die from the effect of dirty water. Another major killer is the use of wood fires for cooking. Often burnt without a chimney, these generate smoke that causes respiratory problems and consequent deaths particularly among women and children.

Precisely because these problems are so horrendous and obvious in their effect, they are the first to be tackled as incomes rise. In domestic energy, for example, we see the natural transitions from wood to kerosene to bottled gas and ultimately natural gas or electricity. As people become better off and more aware of these basic health problems, they naturally seek these cleaner and more convenient options.

Problems of development As countries develop, industrialize and urbanize, a second set of pollution problems emerges. These are typically problems of emissions, often associated with energy use but also including emissions to water. They include sulphur emissions, and other emissions from motor vehicles. Cities such as Bangkok and Beijing provide particularly acute examples of such problems. Modern technology is able to fix most of these kinds of problems, for example, by switching to cleaner forms of fuel or production, or cleaning up the emissions themselves (e.g. by using catalytic converters on cars). These remedies cost money, of course; but developing societies in this category are increasingly able and willing to pay these costs. So we see that as countries get richer, these kinds of problems diminish (Figure 2.6). Indeed, Bill O'Reilly, a former head of the US Environmental Protection Agency, has said that in his view the United States is cleaner now than it was 100 years ago and that effectively these types of problems have been solved in his country.

Figure 2.6: Maximum ozone – Los Angeles basin

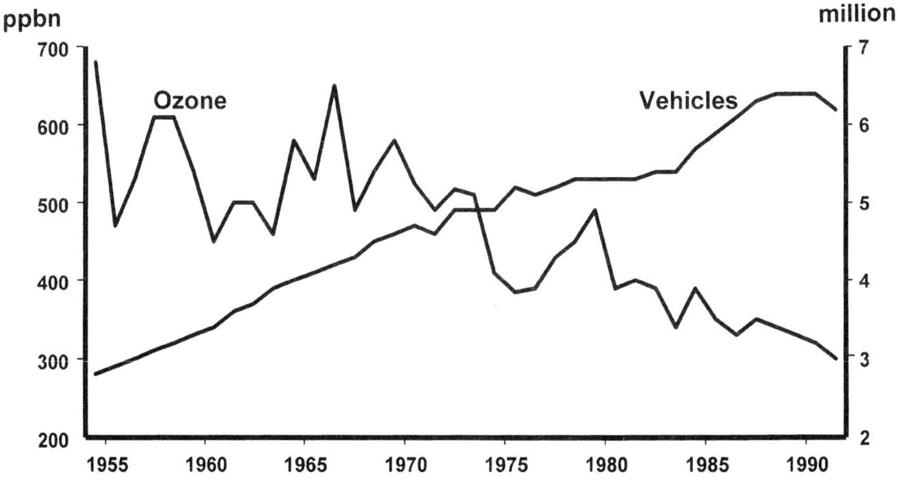

Source: California Air Resources Board.

Problems of affluence There is, however, a third broad category of pollution problems that seem to get worse with increasing wealth. Perhaps the most important, certainly for energy companies, is global climate change, linked to emissions of carbon dioxide and other 'greenhouse gases'. Also in this category are problems of solid waste.

One aspect of this category of problem is that the causes of the pollution and the impact of the pollution are detached from one another. Climate change is again the most obvious and most important example. So, as well as a technological problem, there is a far more complicated institutional problem. How do you manage the amelioration of a problem when the costs and benefits are dispersed both geographically and over time? It is this complexity of resolution that makes sustainable development such a challenge and, of course, will lead to lifetime employment for a large number of scientists, economists, lawyers and bureaucrats.

The social domain

There are many ways of thinking about this domain and many indicators of the well-being of society. Two obvious ones are shown in Figure 2.7: life

Figure 2.7: Life expectancy and education

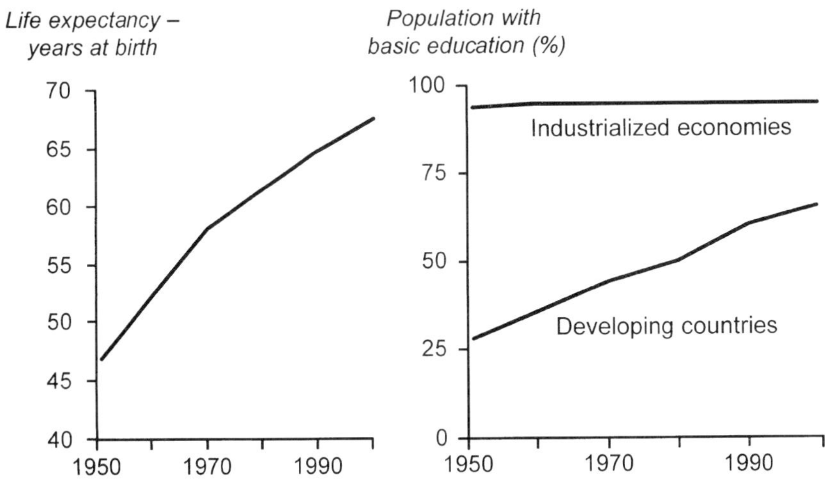

Source: World Resources Institute, 1996.

expectancy and education. Education may be regarded as an input and life expectancy as an outcome. Both have seen significant improvements in recent years. Basic health has seen widespread improvement even in the poorest countries, and many historic scourges have been virtually eradicated. Obviously these gains are linked to advances in wealth and technology.

But we have to include more than these basic human needs in this domain. A useful tool of analysis here is the well-known Maslow 'hierarchy of needs' (Figure 2.8). The poorest and most deprived people seek first of all satisfaction of their most basic needs: food, shelter and warmth. Only then can they focus on other, 'higher' needs. Social needs reflect the fact that most people seek satisfaction from interacting with people outside the immediate family. As a result, we see the creation of a complex web of relationships that underlies whole aspects of economic activity as well as 'civil society'.

Figure 2.8: Maslow's hierarchy of needs

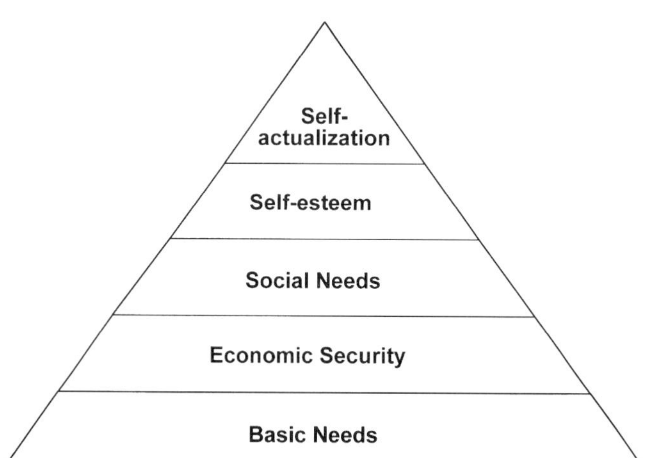

The drivers and linkages of sustainable development now become clear. As economic development allows more and more people to rise through the hierarchy of needs, they both wish and are able to make more choices and to articulate their views. Societies become more complex, and it becomes both impossible and undesirable to restrict the expression of their members' voices through political structures. It is no accident that the richest and best-educated societies become more open and democratic. Americans, of course, would argue that this hierarchy of needs is reflected in the US Declaration of Independence, affirming the individual's right to 'life, liberty and the pursuit of happiness' (the qualifier 'pursuit' shows that not even the most optimistic of constitutionalists believe happiness can be guaranteed by a political process).

It is also no accident that environmental feeling and action are more pronounced in richer and more developed societies. A clean environment becomes desirable and affordable, and more open political and social structures allow people's wishes to be both expressed and acted upon. One way for this to happen is through the proliferation of non-governmental organizations (NGOs): not only those relating to environmental questions but many others ranging over a whole host of humane, social, political and cultural causes.

Globalization

As well as the growing awareness and voice of individuals and institutions, there is another driving force for sustainable development: globalization. We normally think of globalization in terms of trade and investment. Indeed, one of the many ironies in the sustainable development debate is that many of those who are most passionate in their advocacy see globalization as one of the causes of the problems that sustainable development seeks to solve.

Globalization has led to a heightened awareness of environmental and development questions in a number of ways.

- Many environmental problems are global or at least international. Climate change is an obvious example. So too are issues such as the global dispersion of chemical residues and species loss. Even what were thought to be purely local air pollution problems are now seen to have much wider geographical impact. (For example, recent work has discovered that sulphur and other emissions in Northeast Asia can have a significant impact on the west coast of the United States.)

- International actors such as companies, UN agencies and the World Bank are subject to pressure in their base countries and seek to bring best practice to their global operations even when the host governments of other countries in which they operate do not endorse or aspire to the standards of, say, the United States or western Europe.
- The international community is creating a series of standards, regulations and recommendations. For example, the Montreal conventions on ozone-depleting chemicals have mandatory effects. It must be expected that such standards will increasingly develop, not just in the environmental domain but also in the social domain. The UN Declaration on Human Rights is increasingly being invoked to set standards for appropriate political behaviour.

Just as globalization is degrading notions of economic sovereignty, so it is also affecting many long-held tenets of political sovereignty. While these pressures hardly ever originate in an overall drive for sustainability in the holistic sense, the effect is identical.

It is only, therefore, stating the obvious to assert that any company wishing to be a global player will have to take the principles of sustainable development fully into account. The pressures leading to a wider understanding and implementation of sustainable development principles are too deep to be easily reversed or ignored.

Linkages, commonalities and conflicts

All three domains are subject to the same pressures, those coming from the wishes of individuals and organizations and also those of globalization. Sometimes these pressures give rise to a commonality of issues across the domains; sometimes they lead to conflicts. These are illustrated in Figure 2.9.

The economic and environmental domains are the ones often seen to be in greatest conflict. But, as we have seen, many environmental problems can be fixed by more money, and people are usually willing to spend these amounts when they are rich enough. This may also be true of the third category of global problems such as climate change. As realization of the problem and conviction that it is real spread, we may be surprised by the speed at which people accept and press for remedial measures.

Figure 2.9: Commonalities and trade-offs

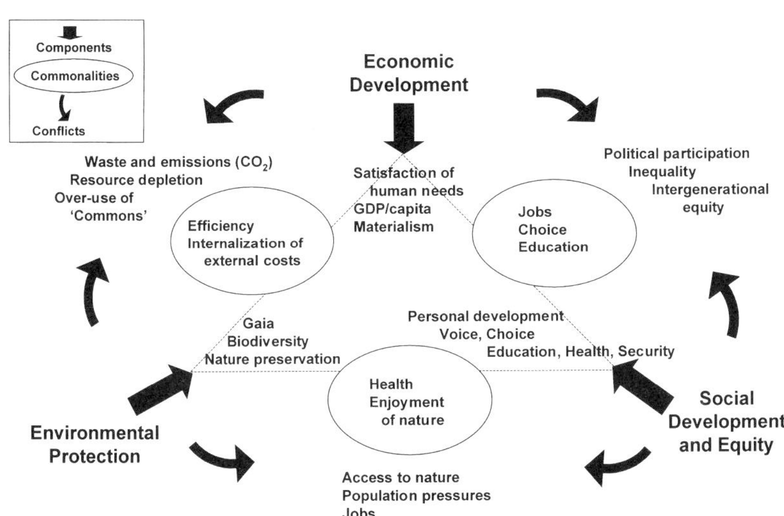

Source: Shell International Ltd.

There is another underlying commonality, which is the abhorrence of waste. In economic systems, waste has a cost. So it comes as no surprise that most industrial pressures are towards ever greater efficiency, generating less waste and, normally, less pollution. Energy is a classic example. Over a long period of time the average efficiency of energy use has improved by about 1 per cent per year, partially offsetting the full increase in demand that economic growth would otherwise cause. There is a double effect here. The costs are more noticeable to heavy users of energy, and we therefore see the greatest drive for efficiency in these industries; but also, as countries get richer, their need for 'stuff', typically industrial goods, saturates and growth is greater in the less resource-intensive service sector.

These effects do not come about if people do not get richer quickly enough, or if new industries are more polluting than those they replace. Poverty is one of the main enemies of sustainable development. Not only does it directly diminish well-being and lock in polluting behaviour, but it is also correlated with higher population growth rates, which lead in turn to higher resource use and pollution.

If the relationships between the economic and environmental domains are the most understood and discussed, the other sides of the triangle of Figure 2.9 also raise important policy issues. For example, on the right-hand side, successful economic development has been shown to require an educated workforce and, of course, supplies better and better jobs, both important indicators of social well-being. On the other hand, these interactions are likely to lead to conflicts, particularly around equity. At issue is not just inequality between rich and poor, but also the whole issue of intergenerational equity. How do we develop policies and measuring systems that balance the needs of people today and of their children and grandchildren tomorrow? The swelling debate around pension provision is a good example of such concerns.

Finally, the third, bottom side of the triangle covers a number of emerging issues. Growing population numbers and people's wishes to see, explore and live in previously pristine parts of the world set up a number of environmental dilemmas. On the positive side, a cleaner environment leads to a healthier as well as a more pleasant world.

This kind of analysis of commonalities and conflict allows us to avoid both the extremes of 'panglossianism', a blind faith that everything will sort itself out, and the doom-laden, end-of-the-world type of messages that emerge from the more gloomy side of the debate. It also presents actors such as companies and governments with a choice: whether to focus on the commonalities – the 'win–win' type of actions – or to grapple with the inherent difficulties of trying to address the conflicts. Society expects companies to find solutions. This is a strong argument in favour of their focusing on areas of commonality, leaving the areas of conflict to be resolved through the more participatory channels of government and politics.

Issues for oil and energy companies

Sustainable development raises a number of issues for oil and other energy companies. To cover them all would require a large volume; the following sections deal with some of the key ones.

Pollution and emissions

Pollution may be caused by emissions from operations and refineries, and in the use of fuels. Pressures to solve these kinds of issues (which tend to be in the middle curve of Figure 2.5) are growing in rich countries. Societies are increasingly able and willing to pay for solutions, which tend to be technical in nature, and expect companies to solve the problem. The days when oil companies vehemently opposed reducing lead in gasoline are long gone. Naturally, there are bound to be disputes about trading off costs and benefits; but is clear that companies are expected to and are planning to minimize their emissions.

Impact on societies

The problems that Shell faced recently in Nigeria highlight another issue, this time located more in the social domain. People have written about the 'curse of oil'. Oil, at least when prices are reasonably high, generates high economic rents. Too often, those in authority struggle to gain control over these rents. Even where the proceeds are spent wisely, large oil revenues tend to distort the economy and make it over-dependent on this single resource. And too often the proceeds are not spent wisely, finishing up in private bank accounts in Switzerland or the Cayman Islands. Oil companies are liable to become guilty by association.

Pressures of globalization will increasingly be brought to bear on oil companies, even where their influence is extremely limited. This is a new dilemma for the companies and is leading to a great deal of thinking. Discussions are taking place with NGOs to explore the limits of possibilities and ground rules are starting to be laid down.

Global climate change

Climate change or global warming is perhaps the biggest sustainable development issue facing energy companies. Fossil fuels when they are burnt emit carbon dioxide (CO_2). Concentrations of CO_2 in the atmosphere have been increasing and are thought to lead to the 'enhanced greenhouse effect'. That is, greenhouse gases such as CO_2 trap the sun's heat in the atmosphere and lead to a gradual warming of the planet. Other gases associated with oil and gas production, particularly methane, are also greenhouse gases.

The great majority of scientists knowledgeable in this area subscribe to the principle that CO_2 emissions will lead to eventual warming and other climate change (see Chapter 3). The science is not yet proven and may never be, but it is idle for energy companies to dismiss the science; no one will care what they say. Shell, with others, believes that the world community should begin to take prudent, precautionary measures (including working towards a better understanding of the science). It supports in principle the Kyoto Protocol signed by 67 governments in 1998.

Of course, this is not the end of the story. The world remains very dependent on oil, coal and gas for its economic well-being and will continue to do so for a very long time. What, therefore, can be done, and how can the oil and gas industry justify its existence or plan for its future?

Responses to the challenge

I would like to make two points in this debate. The first is illustrated in Figure 2.10. What this figure shows is the increase in CO_2 *concentrations* in the atmosphere that would result if all the world's known economic resources of fossil fuels were burnt. Note again the word 'resources': this measure is arrived at by taking proven *reserves* and adding on estimates of what is still to be discovered or retrieved as a result of improved technology.

It can be seen straight away that estimates of resources of conventional oil and gas are rather limited in comparison with unconventional heavy oils (e.g. the Canadian tar sands) and gas (e.g. coal-bed methane). But both pale into insignificance relative to the huge amount of coal.

The horizontal lines are various estimates of what the biosphere can 'safely' cope with. The lower line roughly accords with Greenpeace estimates; the middle one is the EU target (calculated simply as double the pre-industrial level); the upper line is an estimate by the Batelle Institute, an American research institute, of where the costs and benefits of limiting concentrations balance out.

Figure 2.10: Atmospheric concentration for total resource use*

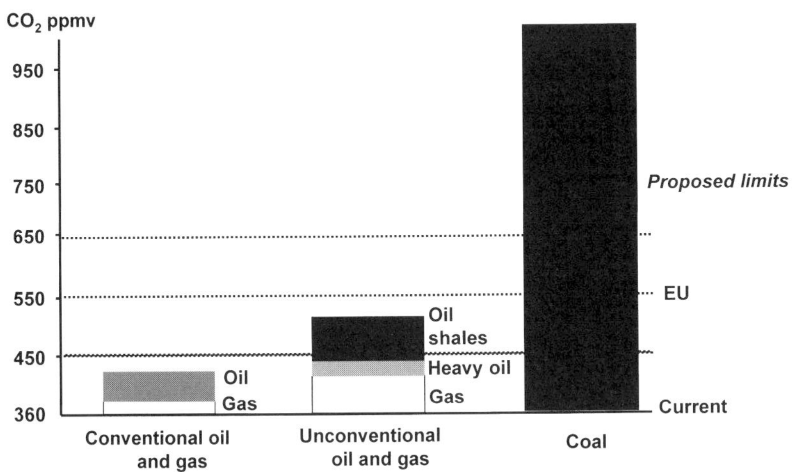

* Proven undiscovered resources at the 50% probability level.
Source: Based on IPCC, Second Assessment Report, 1995; and Charles D. Masters et al., *World Petroleum Assessment and Analysis*, Proceedings of the 14th World Petroleum Congress (New York: John Wiley, 1994).

One thing stands out from this: oil and gas are not the main issue. If we believe the science of climate change and its possible negative consequences, it is imperative that not all the world's resources of coal are burnt.

I would go further. Given the timescales and quantities involved, it is inevitable that the bulk of the world's oil resources indicated in this chart will be used. Oil is wonderful stuff: cheap, easily transportable and with a high energy-to-volume density. There are no real substitutes for its use in transport. Shell's analysis shows that the most severe policies imaginable will not change this fact, even though demand growth may not be as rapid as some believe. This is not bravado, or the defence of the indefensible. Oil is for the time being a vital resource for economic development. Without it, crops cannot get to market, ambulances to hospital, or people to work (and leisure). Hence its use is vital for satisfaction of the economic domain; and, as we see, it is not catastrophic in climate change terms. Naturally, this does not remove the need to improve the efficiency of its use.

I also believe that gas is a vital transition fuel. The world's energy system during the course of the twentieth century has been steadily 'decarbonizing'. As we have shifted from an almost complete dependence on coal to increased use of oil and, later, gas, the amount of carbon emitted per unit of energy has decreased. At one time nuclear power was seen as the ultimate step on this journey, capable of providing abundant and clean energy too cheap to meter. This dream lies shattered, but the decarbonization trend continues. A major factor today is the increased use of natural gas, particularly in power generation. Not only does gas have a lower carbon-to-energy ratio than oil or (particularly) coal, but its efficiency in power generation is substantially higher than that of coal because of the combined-cycle gas turbine (CCGT) technology. CCGT has other advantages as well. Low initial capital cost, quick building times and smaller generating plant size have made it the economic plant of choice; it also produces lower local emissions (of sulphur and nitrous oxides) and less carbon. No wonder that, where it is available, gas is the fuel of choice today. This trend is only reinforced by the perceived problems of nuclear power. Shell sees gas as an essential part of the transition to a sustainable energy world.

This is not all. Some who worry about the future of the energy world have postulated that when all the oil and gas have been used up, the world would then have to turn to its vast coal reserves and heat up the planet. This is the basis of the International Panel on Climate Change's 'business as usual' scenario produced in 1992 (see Chapter 3). In Shell's view, this scenario is implausible. It would have policy-makers in, say, 2050, in a world where climate change has been demonstrated to be happening (otherwise there is no point to the scenario), telling our grandchildren that, now oil and gas have run out, they have to go into the mines and dig out more and more coal, even though it will make the planet hotter yet.

Shell's own long-term energy scenarios tell a different story. In one version, which we call 'Dematerialization', there will be dramatic improvement in energy efficiency, and a strong movement of wealth creation away from heavy industry towards services, reducing the impact of energy use on the environment. We can see the first fruits, for example, of work being done to produce much more efficient motor vehicles. The chairman of Ford commented in 1995 that new technologies such as advanced electronics, ultra-light materials and computer-aided design could change cars more radically in the next 10–20 years than in the last 100. In this scenario early introduction of such cars, encouraged by the

Figure 2.11: The new generation of motor vehicles

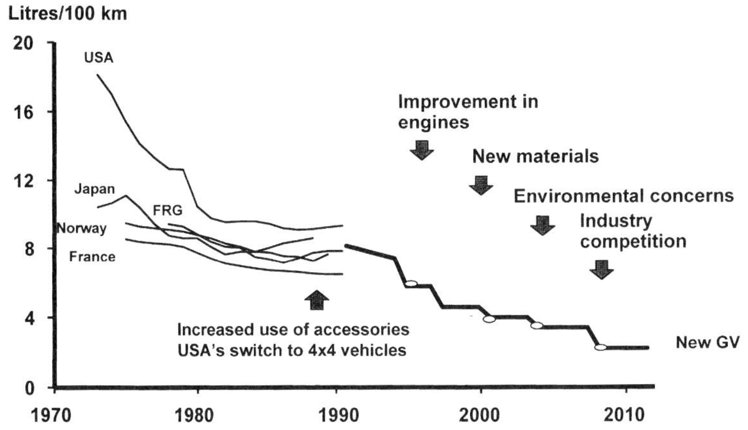

Source: Shell International Ltd.

US PNGV (Programme for the Next Generation Vehicle), results in dramatic improvements in fuel use (Figure 2.11).

In another scenario, called 'Sustained Growth', a continuation of recent downward trends in the cost of new renewable energy makes these energy sources more and more competitive (Figure 2.12). As this happens, these energy forms find new niches and as demand for them increases, production increases result in further cost reductions. This is the well-known 'learn-

Figure 2.12: Newcomers in electricity

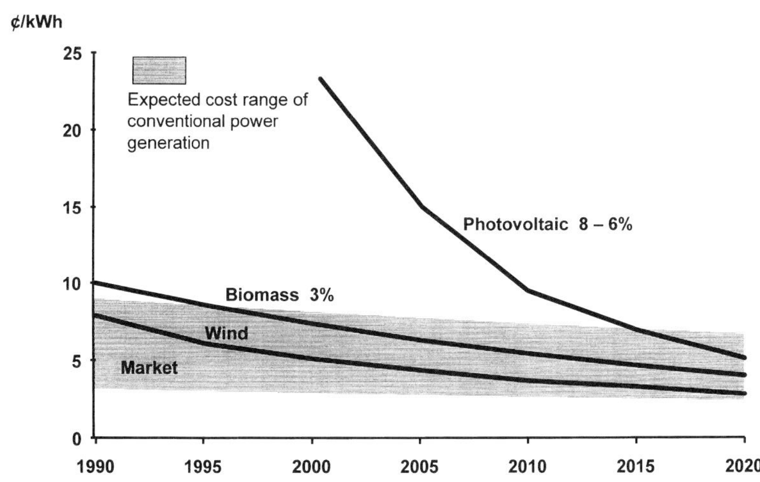

Source: Shell International Ltd.

Figure 2.13: Shell's 'Sustained Growth' scenario

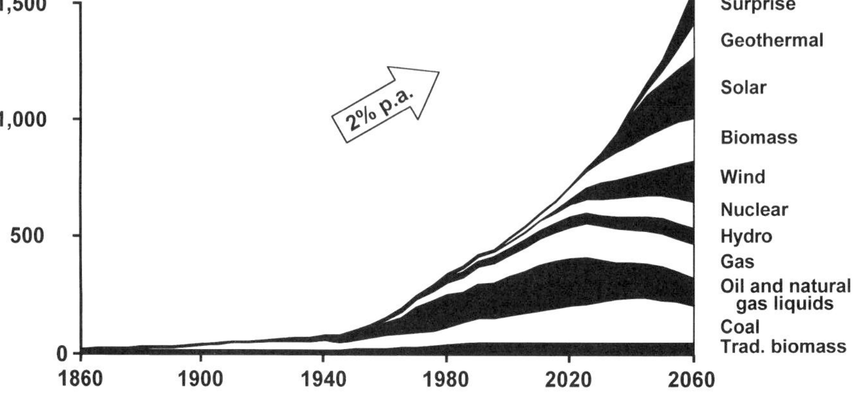

Source: Shell International Ltd.

ing curve'. Hence by 2020/2030 these new (and carbon-free) energy sources will be ready to take over from oil and gas as they peak and start to decline. This scenario is illustrated in Figure 2.13. In this world non-fossil fuels supply more than half of global energy needs from about the middle of the twenty-first century and continue to grow in importance thereafter.

Interestingly, the path of carbon emissions in these two worlds is not dissimilar. In both of Shell's long-term scenarios, CO_2 emissions peak before the middle of the century, in strong contrast to the IPCC scenario (IS92a) mentioned above. Shell's scenarios do not reflect strong government intervention in energy markets. Were such intervention to occur – in response to the Kyoto Protocol, for example, or for other reasons – then that would tend to accelerate the trends of sustained growth and dematerialization. The fourth line in Figure 2.14 shows the outcome of a very stringent set of actions described in a World Energy Council scenario.

These kinds of scenarios, together with the analysis in Figure 2.10, have led Shell to a clear set of policy recommendations and the development of its own strategies:

- Oil and gas are a necessary part of the transition to a low-carbon energy world.

Figure 2.14: CO_2 emission – fossil fuels

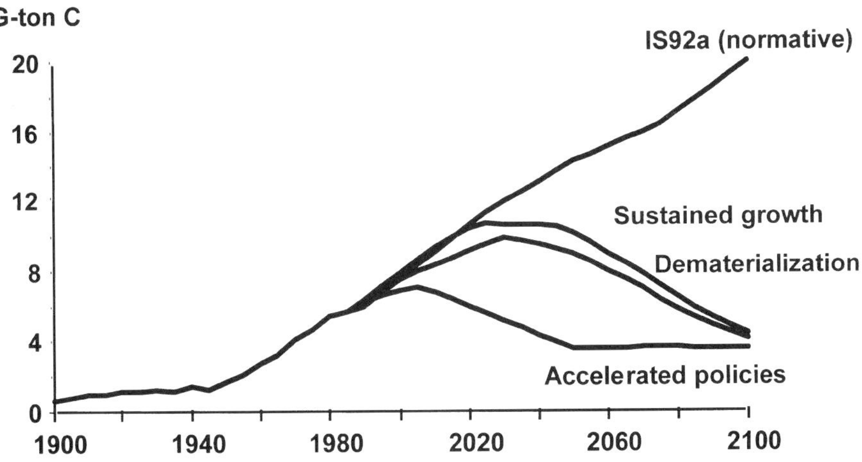

Source: Shell International Ltd.

- Gas in particular will be called upon to play a major role over the next 20–30 years, to ensure electricity is produced as cleanly and efficiently as possible. The role of gas will be further accentuated if nuclear power is unable to re-establish itself as an economic and environmentally acceptable energy source. Shell is the world's second largest gas company (after Russia's Gazprom) and intends to play a full role in making this happen. Governments should try to ensure that moves to gas are not blocked by subsidizing coal.
- Renewables will gradually penetrate energy markets and increase their market share. Shell is not alone in seeing that this presents a long-term business opportunity and is taking steps to position itself in those areas where it thinks it can be competitive and create value. It would be in line with the precautionary principle approach to climate change for governments to encourage renewables. Given the pronounced skew of existing government energy R&D programmes to coal and fission and fusion, this could be done without having to increase overall spending.
- Economic instruments should be favoured over regulation because they are more likely to result in low-cost, efficient solutions. Carbon taxes have merit in theory. In practice, however, so many exclusions seem to appear when legislation is proposed that they risk ending up

as nothing more than glorified gasoline taxes. This is unjustified and also useless. Moves towards trading systems for CO_2 emissions, on the lines of the US regime for sulphur dioxide emissions trading and exchange, seem to have more promise. Private companies like Shell, as well as governments, are already thinking about how to operate such systems.

The above discussion should have made apparent some of the thinking that lies behind recent responses by Shell and other companies to the challenges of sustainable development. I have touched here upon only a few of the key issues that impact on the oil and gas businesses. Every business will have its own particular areas of concern, ranging from common issues of pollution to specific resource use questions and the impact of new technologies (e.g. genetic engineering). Companies take these challenges seriously because they recognize the logic that makes sustainable development such a powerful concept for government and international action. They see that as this logic works through, it will change the nature of the markets into which they sell and the wishes of their customers.

Moreover companies, particularly large global companies, see that they have an important role to play. The impacts of some of the issues raised above – global warming is a typical example – are only seen over long time periods. Yet many institutions are governed by much shorter timeframes (Figure 2.15). Democratic governments have to live with the constraints imposed by the electoral cycle. If companies want a longer-term stable business environment, they have a responsibility to do what they can to bring it about. Increasingly society will exert pressure on companies – through the behaviour of customers in the market-place, government bodies or other NGO lobbying activity – to exercise that responsibility. Far-sighted companies see a natural merger of responsibility and self-interest when they begin to accept and respond to the challenges of sustainable development.

In Shell's case, the Brent Spar and Nigeria episodes have been a painful way to internalize some of the issues. As we have tried to learn the lessons and to understand the feelings and thoughts of the wider world, we have changed our own mental model or myth (Figure 2.16). Before, we saw ourselves as an agent under enormous pressure from different

Figure 2.15: Mismatched timetables

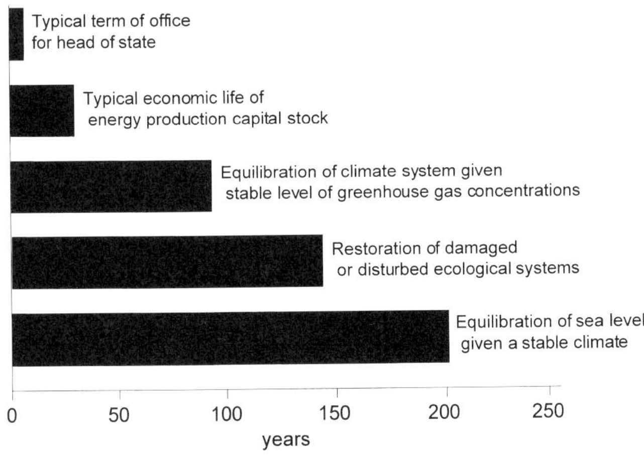

Source: Adapted from IPCC, *Second Assessment Report*, 1995.

stakeholders, trying to balance each against various others and perpetually in reactive mode. Seeing ourselves as an integral part of society with clear duties and responsibilities has led to a major shift in perception and has allowed a clearer view of what can and should be done.

What of the wider oil and gas industry? Too often it has been defensive, identifying its best interest in the attempt to deny the importance and even existence of global warming. OPEC has presented itself as a victim of a

Figure 2.16: Mental models

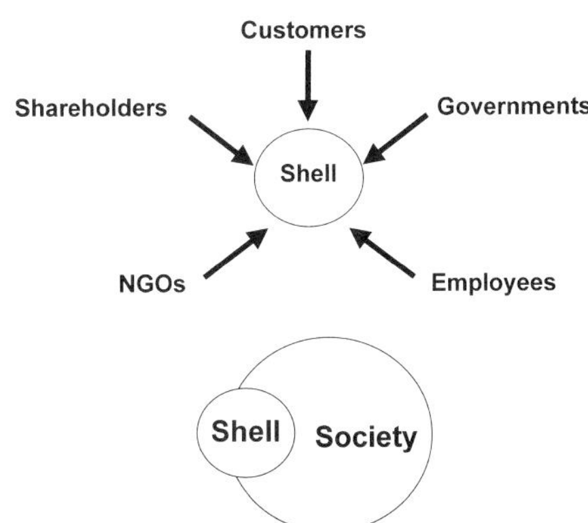

Source: Shell International Ltd.

hostile conspiracy. By doing so, it has sidelined itself in the debate and devalued some of its legitimate arguments.

My personal view is that the oil industry and oil producers have a much more positive story to tell. Abundant and rather cheap oil has been, is and will be for at least the next two or three decades a key and irreplaceable ingredient of sustainable development. It is essential to economic growth, itself a key pillar of sustainable development, and at the basic level substitutes for far more dirty polluting energy sources (e.g. open wood fires). Also, as argued above, gas is the key transition fuel to a more sustainable energy system over the next few decades.

These messages need to be expressed strongly, for there is much misunderstanding among governments and policy-makers. Furthermore, by accepting the argument for rational economic measures that address global climate change, the oil and gas community is much better able to attack the attempts to penalize their interests unfairly – for example, through moves by governments to dramatically increase gasoline and other oil taxes. Of course, governments have the right to tax what they want; but they should not be allowed to dress such measures up as environmental charges. In Europe such taxes are far in excess of realistic environmental costs, and OPEC and other bodies have the right to try to expose the hypocrisy that underlies them.

Also, as we begin to get the point across that oil and gas are part of the solution, we should try to prevent people lumping oil and gas together with coal in the indiscriminate use of the term 'fossil fuels' which puts us all in the dock together. Coal use will not stop overnight, and in places such as China and India will doubtless go on growing. But oil and, particularly, gas offer a better solution.

Conclusions

The ideas that underlie sustainable development will increase in importance, driven by people's growing awareness and wishes on the one hand and the more impersonal forces of globalization on the other.

There are many commonalities among the three domains of sustainable development – for example, wealth creation, efficiency, environmental protection, education and the expression of choice. Actors can focus on these commonalities to seek win–win results. But the conflicts are also real. Dealing with global warming concerns is almost certain to have a real economic cost. Resolution of these conflicts will be messy, not least because the causes and their effects can happen in very different timeframes. The world community and individual countries have shown they can tackle these issues; the Montreal Convention on ozone depletion

Figure 2.17: The triple bottom line

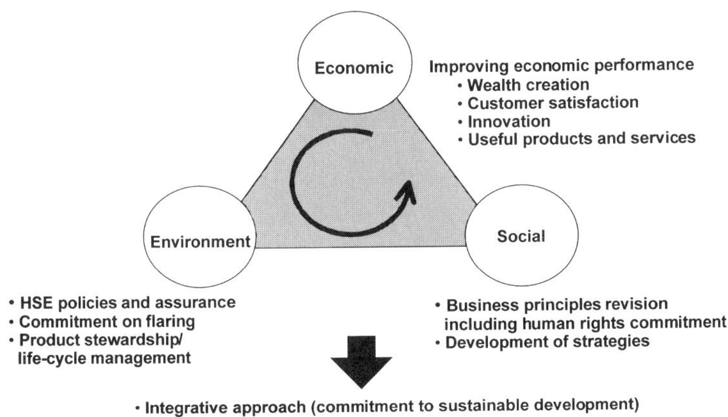

Source: Shell International Ltd.

is a good example. But there is much to do and the corporate sector has a duty to involve itself in seeking acceptable trade-offs.

Finally, Shell believes that the oil and gas industries are part of the solution, particularly in the transition to a sustainable energy world. We have to convince many sceptics that this is so. As we have thought about sustainable development we have expressed our vision and accountabilities in the concept of the triple bottom line, corresponding to the three domains (Figure 2.17). Already we see some of the benefits: the emergence of new customer-driven value propositions (e.g. renewable energy); a framework for developing environmental and social policies; and, very importantly, a basis for participating in the public debates. We will continue to develop our understanding and responses to what will continue to be a key global concept in the twenty-first century.

3 Global warming in an international context

Michael Grubb

One strikingly new issue to emerge into the international geopolitics of energy since the 1970s has been concern about man's impact on the climate. It is an issue which many people are still struggling to place in the international political and energy policy scenes; and yet it is one which could have the most profound consequences for the future of energy and which bears upon many of the other issues identified in this volume. In this paper I discuss the origins and evolution of the climate change issue, beginning with some scientific background and continuing with discussion of the economics and the political development of the international response.

Scientific background

The idea of climate change is not a new one at all. There was a scientific paper published in 1832, more than 150 years ago, which said that the effect of certain gases in the atmosphere must be to trap heat near the surface, and this explains why the surface of the earth is much warmer than it would be if there were no atmosphere. That is an uncontested fact: if we did not have the greenhouse effect, if we did not have greenhouse gases, we could not live on the planet. The most important of those greenhouse gases are water vapour and carbon dioxide. Both of those act quite powerfully to trap reradiated heat from the sun in the lower atmosphere, keeping the earth's surface warm. And more than a hundred years ago another scientific paper speculated that the age of coal, the industrial revolution built on coal, would lead to an increase in carbon dioxide and thereby to a warming of the planet's surface. But it has to be said that very little happened in the next sixty years in terms of thinking about this issue. It only began to come to life again as a subject of scientific study in the 1960s, when various works began trying to quantify the effect of emissions of carbon dioxide on the atmosphere and thereby global warming.

The key developments in which the degree of scientific concern began to penetrate government circles took place during the 1980s, when the UN Environment Programme and the World Meteorological Organization convened a series of scientific conferences which began to build a consensus upon the nature of this problem, against a background of rising popular concern about environmental issues. In November 1988 a meeting of 35 governments in Geneva established the Intergovernmental Panel on Climate Change, under the auspices of UNEP and the WMO. All governments were invited to join, and the IPCC expanded to almost universal participation. It is the reports of the panel that provide the scientific underpinning for the diplomatic processes of the UN Framework Convention on Climate Change.

The purpose of the IPCC is to provide governments with authoritative assessments of the state of knowledge concerning climate change. In the ten years since it was established, it has evolved into what is probably the most extensive and carefully constructed intergovernmental advisory process ever established in international relations. Governments nominate experts with an established record of research and publication on relevant topics. From this pool, writing teams within each of three working groups are selected by the bureau of governmental representatives for their level and breadth of expertise, range of views and geographical balance. The texts they produce are subjected to widespread peer review involving hundreds of individuals and groups worldwide.

The IPCC produced its first report in 1990. The findings of Working Group I (Science) probably had the greatest impact, with its key conclusion that rising concentrations of carbon dioxide and other greenhouse gases in the atmosphere were caused by human activities and would cause global temperatures to rise, with accompanying climatic changes. As the IPCC sought to explain, certain scientific fundamentals are undisputed. Greenhouse gases trap heat near the earth's surface. The planet would be much, much colder without them. The concentration of greenhouse gases is rising, primarily owing to human activities. That increase is likely to lead to warming of the earth's surface on average, though with important regional variations. These are the simple fundamentals on which all could agree.

Beyond that point the picture gets more complex, principally because of the huge variety of additional influences upon the global atmosphere and temperature. These include natural feedbacks – changing global temperature itself changes the water cycle and associated cloud and ice cover, for example. They also include many extraneous influences, both natural (ocean current oscillations, sunspot changes) and human (sulphate and other emissions, and land use changes). Disentangling these influences to predict the magnitude and consequences of human-induced climate change is a challenge of an altogether different order.

The first IPCC report produced some 'best-guess' projections of climate change, and agreed that the earth's surface had warmed, but was not able to say whether this was due to human-generated emissions. That changed five years later with the IPCC's *Second Assessment Report*. In what was politically the most important finding, the scientists concluded that 'the observed warming trend is unlikely to be entirely natural in origin'. The inclusion of aerosols, improved modelling (including of natural variability) and better data on the distribution of temperature changes brought model-simulated changes in various dimensions much closer to those actually observed. On the basis of such 'pattern matching' studies, governments accepted that 'the balance of evidence suggests that there is a discernible human influence on global climate'. Other conclusions are summarized in Box 3.1.

On the IPCC's central emissions projections, by the end of the next century global mean surface temperature is likely to rise by about 2°C, with a range of uncertainty of 1–3.5°C, and to continue rising for some decades thereafter even if greenhouse gas concentrations are stabilized by then. The numbers may appear small – we go through much more than that in terms of the change outside between night and daytime – but in terms of a global average it is a very important indicator. The last ice age was 4–7°C colder than today, and for most of the last 10,000 years – during which human civilization has developed – the global temperature has oscillated by less than half a degree one way or the other.

The IPCC identified various other specific effects. Sea level would rise, with a mid-range estimate of 50 cm by 2100 (range 15–95 cm), and would continue rising for centuries thereafter. Various regional changes

Box 3.1: IPCC *Second Assessment Report*: summary of key elements for policy-makers

- Greenhouse gas concentrations have continued to increase as a result of human activities.
- Recent years have been among the warmest since at least 1860; the twentieth century has been warmer than any since reliable estimation possible (probably warmer than any since beginning of last ice age).
- The ability of climate models to simulate observed events and trends has improved.
- Emerging evidence points towards detectable human influence on climate.
- Not known whether climatic variability and frequency of extreme events has changed on global scale, but significant changes in certain regions.
- Best-guess changes for central (IS92) scenario by year 2100:
  - global average temperature rise: 2°C; night-time temperatures are expected to rise more than daytime;
  - sea-level rise: 50cm, and would continue rising for centuries even if the atmosphere were to be stabilized in 2100.
- Large and rapid climatic changes have occurred in the past, although the last 10,000 years have been relatively stable.
- There are continuing major uncertainties, with most emphasis upon those concerning the behaviour of ocean currents.

in rainfall and storm patterns would be expected, though the detailed nature of these changes was harder to predict. Certain species would move or become extinct, and tropical diseases might spread more widely. Thus there would be significant and often adverse impacts on many ecological systems, potentially including food supplies and water resources, and on human health. In some cases, those changes would be irreversible; and the developing countries, and particularly small island countries, would typically be more vulnerable (developing countries are

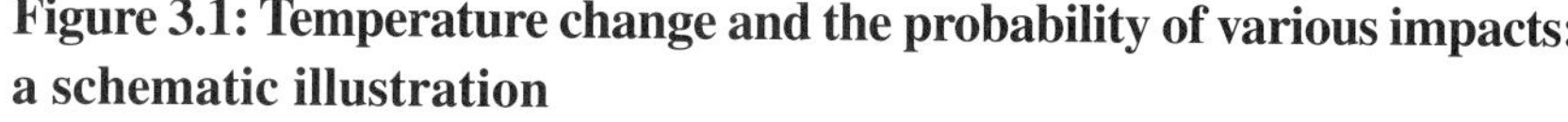

Figure 3.1: Temperature change and the probability of various impacts: a schematic illustration

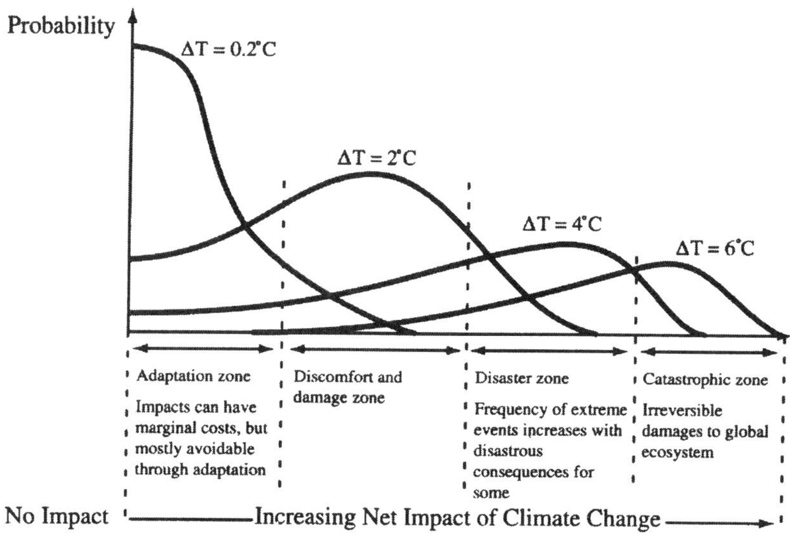

Source: Parikh et al. (1995), presented in Intergovernmental Panel on Climate Change, *Climate Change 1995: Economic and Social Dimensions* (Cambridge: CUP, 1996), Chapter 2, p. 60.

predominantly in regions which are already warmer and often dryer, and the social and economic infrastructure is not strong enough to adapt easily to potential climate change).

Many uncertainties remain, and there is a tendency for environmental groups to try to simplify things. The fact is that we do not really know what will happen with climate change; but most of those involved in studying the possibilities would accept something like what is depicted in Figure 3.1, which tries to illustrate the probability of certain kinds of effects. Small temperature changes are easy to adjust to; larger temperature changes bring more uncertainty, with higher risks of more extreme events; and very big changes increase the chances of potentially catastrophic changes, in ocean currents, for example, which have a huge impact upon regional climates. That is one way of thinking about why governmental and political action to reduce emissions is already beginning even though many of the details remain uncertain.

The political response

At the Second World Climate Conference of November 1990, governments accepted the IPCC's *First Assessment Report* and called upon the UN to launch negotiations on a Framework Convention. The negotiations formally commenced in February 1991 and proceeded intensively and – by the standards of international negotiations – quickly, to have a convention ready for signing at the Rio Earth Summit in June 1992. The Convention establishes a set of principles, institutional procedures, and an interim goal for the industrialized countries to establish leadership of global efforts by aiming to return their emissions to 1990 levels by 2000.

Following these intensive efforts, the political profile of the climate change issue collapsed in exhaustion. The public was bored with the story and the media, if they bothered to pursue it at all, were more interested in running sceptical challenges to the new-found and frail consensus. Yet beneath the surface the institutional mechanisms set in train by the IPCC and the Framework Convention continued their work, exploring the issue and its implications, and slowly gathering new force. The Convention, which by the end of the Earth Summit had already attracted more signatories than the GATT, was ratified and entered into force with remarkable speed. Governments embarked upon the complex process of trying to develop policies in line with the commitments made under the Convention, and to submit national reports on the measures taken and on emissions projections.

In March–April 1995, the First Conference of the Parties to the Convention gathered in Berlin to consider the next steps, and in particular to carry out their mandated task of examining the adequacy of the existing commitments. Faced with additional scientific evidence and still growing emissions, it was impossible to reach any conclusion other than that the commitments were not adequate. Hence a new round of negotiations was launched under the Berlin Mandate, to develop by late 1997 new commitments on 'quantified limitation and reduction objectives' (emissions targets) and on policies and measures (Box 3.2).

Probably the defining point of these negotiations was the US statement to the Second Conference of the Parties. The US negotiator first rounded on the US industrial critics of the IPCC's *Second Assessment Report*:

Box 3.2: The 'Berlin Mandate' for the Kyoto negotiations

'1. The First Conference of Parties, having reviewed [the commitments] and concluded that these are not adequate, agrees to begin a process to enable it to take appropriate action for the period beyond 2000, including the strengthening of the commitments of Annex I (Industrialized) Parties ...

'2. The process will, *inter alia*:

'(a) aim, as the priority in the process of strengthening the commitments [for industrialized countries], both

- to elaborate policies and measures, as well as
- to set quantified limitation and reduction objectives within specified timeframes, such as 2005, 2010 and 2020, for their anthropogenic emissions by sources and removals by sinks of greenhouse gases not controlled by the Montreal Protocol taking into account the differences in starting points and approaches, economic structures and resource bases, the need to maintain strong and sustainable economic growth ...

'(b) not introduce any new commitments for [developing countries] but reaffirm [their] existing commitments and continue to advance the implementation of these commitments in order to achieve sustainable development ...'

> The IPCC's efforts serve as the foundation for international concern and the clear warnings about current trends are the basis for the sense of urgency within my government. We are not swayed by and strongly object [to] the recent allegations about the integrity of the IPCC's conclusions ... raised by naysayers and special interests bent on belittling, attacking and obfuscating climate change science ... let me make clear the US view: the science calls upon us to take urgent action.

He then went on to declare:

> Our analysis and consideration of this issue to date have led us to certain conclusions ... Sound policies pursued in the near term will allow us to avoid the prospect of truly draconian and economically disruptive policies in the future ... denial and delay will only make our

> economies vulnerable in the future.
>
> The US will seek market-based solutions that are flexible and cost-effective ... The US recommends that future negotiations focus on an agreement that sets a realistic, verifiable and binding medium-term emissions target ... met through maximum flexibility in the selection of implementation measures, including the use of reliable activities implemented jointly, and trading mechanisms around the world.

With the basic US position clarified, the next 18 months was a period of frenzied negotiations which became highly political and contentious, but which did ultimately culminate in agreement on the protocol in Kyoto in December 1997. The protocol that emerged was largely shaped by the US administration, particularly in respect of industrialized country commitments and the various mechanisms for 'flexibility' that surround them.

The Kyoto Protocol

The core commitments of the Protocol are summarized in Table 3.1.[1] It establishes legally binding constraints on emissions of greenhouse gases from each industrialized country listed in Annex 1 of the convention. That includes all of the OECD except Turkey and the new entrants of Mexico and Korea. It includes eastern Europe and, of the former Soviet Union, Russia, Ukraine and the Baltic countries; but it does not include the Central Asian countries. Emissions are defined over a commitment period which is centred on a point more than ten years away, 2008–2012. The commitments are expressed in relation to emissions during a base year (generally 1990, which in most respects is taken as the base year for the international climate change negotiations).[2] A different base year is allowed for some of the transitional economies of central Europe, and for industrial trace gases. The specific commitments are set out in Table 3.1.

[1] For a detailed discussion of the Protocol, its negotiation and provisions, see M. Grubb, C. Vrolijk and D. Brack, *The Kyoto Protocol: A Guide and Assessment* (London: RIIA/Earthscan, 1999).

[2] It was in 1990 that the IPCC produced its *First Assessment Report* and the ministerial segment of the Second World Climate Conference agreed to launch the negotiations that resulted in the UNFCCC at Rio in 1992. This year has subsequently become the focus of national inventory assessment for greenhouse gas emissions, ensuring that 1990 remained the base year for commitments, rather than an average over 1988–92 which was proposed in the month before Kyoto to complement the five-year average basis for commitments.

Table 3.1: Emissions and commitments in the Kyoto Protocol

Country	*Commitment (% change from base year†)*	*1995 emissions (% change from base†)*	*Other GHGs (% 1990 global warming potential)*
Australia	+8	+4.9	51.9
Canada	–6	+6.3	18.4
European Union*	–8	–2.3	20.3
Iceland	+10	+4.7	25.6
Japan	–6	+6.5	5.8
Liechtenstein	–8	n.a.	n.a.
Monaco	–8	+82	0
New Zealand	0	+2.3	68.7
Norway	+1	+4.3	34.5
Switzerland	–8	–1.6	17.8
United States	–7	+4.3	15.2
Economies in transition			
Bulgaria	–8	–28.8	28.8
Croatia	–5	n.a.	n.a.
Czech Republic	–8	–19.1	13.9
Estonia	–8	–35.7	16.6
Hungary	–6	–24.1	17.7
Latvia	–8	–45.7	16.9
Lithuania	–8	n.a.	n.a.
Poland	–6	–19.1	14.6
Romania	–8	–27.12	30.5
Russia	0	–22.7	22.4
Slovak Republic	–8	–15.8	17.8
Slovenia	–8	+6.1	n.a.
Ukraine	0	–35.6	n.a.

* The 15 EU member states are each listed as having a target of –8%. These targets were subsequently redistributed under the 'bubbling' provisions of Article 4; see text below and Table 3.2.

† Base year is generally 1990, but varies for some economies in transition. For full details see Grubb et al., *The Kyoto Protocol: A Guide and Assessment*, Table 4.1.

Coverage and time periods

The total of the commitments adds up to countries reducing their greenhouse gases by at least 5 per cent below 1990 levels. The commitment is defined in terms of the basket of the major greenhouse gases, which are aggregated for the purposes of the commitment. Some element of 'sinks' may be included: these are defined as actions that absorb carbon dioxide from the atmosphere, resulting from direct human-induced land-use change and forestry activities since 1990.

The commitments for 2008–12 serve as the engine for almost everything else in the Protocol. There is no earlier commitment period, though Article 3.2 states that 'Each Party shall, by 2005, have made demonstrable progress in achieving its commitments under this Protocol.' The Protocol makes reference to a second commitment period to succeed the first, and stipulates that negotiations to define corresponding commitments shall start no later than 2005. Emissions reductions obtained over and above commitments in the first period may be 'banked' against commitments in the subsequent period.

The most prominent commitments are for the European Union (–8 per cent), the United States (–7 per cent), Japan (–6 per cent) and Russia (maintenance at 1990 levels). Some smaller countries gained greater derogations. The Protocol allows a group of countries to redistribute their commitments within a collective cap, and the EU subsequently defined targets for its member states using this provision (see Table 3.2).

The 'assigned amounts' for Germany and the UK are relatively ambitious, as are those for Japan and Canada in the Protocol itself. Perhaps the

Table 3.2: The internal distribution of the EU 'bubble'

Country	*Internal commitment (%)*
Austria	–13
Belgium	–7.5
Denmark	–21
Finland	0
France	0
Germany	–21
Greece	+25
Ireland	+13
Italy	–6.5
Luxembourg	–28
Netherlands	–6
Portugal	+27
Spain	+15
Sweden	+4
United Kingdom	–12.5

defining challenge, however, is the target of 7 per cent below 1990 levels for the United States. On the surface that is a very big commitment – especially in view of the fact that US emissions in 1997 were about 10 per cent above 1990 levels. However, I referred above to flexible measures on which the US insisted, and those measures were carried forward in the Protocol. These include various mechanisms by which countries can swap commitments or take action abroad.

Broadly, the concepts are quite straightforward. *Emissions trading* allows two countries to 'trade' parts of their permitted emissions under the Protocol. The Convention establishes that there shall be emissions trading between industrialized countries that have accepted caps on their emissions, though detailed rules have yet to be agreed.

Joint implementation enables emissions savings arising from cross-border investments between Annex I Parties to be transferred between them. If such investments are sanctioned by the governments of the participating industries – including an estimate of emissions saved by the investment, as compared with what would otherwise have been emitted – the agreed emission savings become equivalent to an international emissions trade, being deducted from the allowed emissions of the host country, and added to the allowed emissions of the investing country.

In addition to these two mechanisms for flexibility between Annex I Parties, the Protocol establishes a 'Clean Development Mechanism' which, in principle, enables activities similar to joint implementation to proceed with developing countries.

In other words, then, the United States is not necessarily committed to a 7 per cent reduction; it is committed to an allowance which starts at –7 per cent, but which is spread across several gases, can be offset against sinks, and can be transferred internationally.

Overall, therefore, the Protocol is a very flexible instrument. Countries can choose how they will meet emission targets, internally or by international transfers. Ultimately, this could lead to marketization of the environmental problem, by transferring this structure of intergovernmental exchanges to industries. This is very much what the United States is hoping for, and many governments now accept that this will happen. Governments will control their own emissions in part through a domestic ystem of emissions permits. In other words, they will give companies a ermit to emit a certain amount, and those companies may trade both ith other companies within the country and internationally. Thus, the companies themselves will have a right to emit a certain amount and will be able to do a deal with a company in some other part of the world that has a similar commitment.

This should be a much more efficient mechanism than a provision confined to the governmental level, because companies can hunt out where they think it will be cheapest to limit emissions; it also means that companies can see the value of carbon dioxide limitation on their balance sheet. So the economic incentives will change. Companies will have emissions permits as part of their balance sheet: if they hold emissions permits, they have a certain value; if they sell them, it generates cash for the company. Thus this system introduces the environmental problem onto the company balance sheet. Clearly, if the price of permits is high and the cost of limitation is high, then the value of reducing emissions also becomes high and visible on the balance sheet.

There is, of course, quite a substantial international intergovernmental impact as well, because corporate trading in permits may result in aggregate national emissions turning out to be very different from the numbers that have been agreed and publicized.

Other aspects of the Kyoto Protocol

Several other aspects of the Protocol deserve mention. First, it does specify certain policies and measures. Governments are required to implement various general policies and measures, such as research and development, and progressive reduction or phasing out of the market imperfections that run counter to the Protocol's objectives (notably, though these are not mentioned specifically, German coal subsidies).

One clause is of particular interest to energy-exporting countries. A defining tension in the negotiations on climate change, ever since their inception, has been that between the concern of some developing countries (especially members of AOSIS, the Alliance of Small Island States) about the adverse impacts of climate change on them, and the contrasting concerns of energy exporters (especially OPEC members) about the adverse economic impacts of response measures. Both sets of concerns were in fact shared more widely among the developing countries. A major political success of OPEC was to align its fears procedurally alongside those of other 'vulnerable groups' listed in the Convention. A key clause in the Protocol's Article 3 on commitments is the final paragraph:

> Each Party included in Annex I shall strive to implement the commitments ... in such a way as to minimize adverse social, environmental and economic impacts on developing country Parties, particularly those identified in Article 4, paragraphs 8 and 9, of the Convention ... [The Conference/Meeting of Parties shall] ... consider what actions are necessary to minimize the adverse effects of climate change and/or the impacts of response measures on Parties referred to in those paragraphs. Among the issues to be considered shall be the establishment of funding, insurance and transfer of technology.

This reinforces, and is more specific than, the similar provision in Article 2 on policies and measures. In addition, a decision appended to the Protocol at Kyoto launches a specific fast-track process 'to identify and determine actions necessary to meet the specific needs of developing countries', referring to the articles on vulnerable groups. These were the key provisions which relieved OPEC members' concerns sufficiently to allow them to accept the Protocol.

The paragraph thus certainly establishes the basis for subsequent negotiation on measures on 'funding, insurance and transfer of technology' relating to vulnerable groups. Arguably it could also be used to challenge abatement measures that OPEC consider discriminatory against international oil. Since the political pressure within Annex I countries will be to adopt measures which tend to protect domestic interests (as with German coal subsidies) at the expense of imported fuels, this could prove to be an important provision.[3]

The Protocol, although mostly concerned with industrialized country commitments, does have some significant provisions relating to developing countries, one of which is the Clean Development Mechanism. This is

[3] Potentially this could set the stage for an enduring struggle pitching the coal industries within Annex I countries against the international oil business, with Annex I governments at the interface. In the short term this will be limited by the complexity of energy trade flows (which include considerable non-OPEC oil trade as well as growing international coal trade); nevertheless, the potential for developing country exporters to delve into the impacts of Annex I energy policies is considerable and could have profound long-run implications.

an area that will be subject to major negotiations in the future, because all sorts of questions arise, for instance about the quantities of emissions saved, who gets the crediting, or the conditions for ensuring sustainable development, and the mechanism will be subject to an intergovernmental executive body of control. For the first time also, Article 12 contains an explicit legal obligation to help vulnerable countries meet the cost of adapting to climate change. This will generate some revenue to help countries to adapt. Developing countries are also required to go a little further in terms of national programmes on climate change, and to take measures promoting technology transfer. Industrialized countries have to promote technology transfer, and the Protocol reaffirms the role of the Global Environment Facility to provide some funding for this as well as for broader environmental efforts in developing countries.

The impact of the Protocol

In 1995, industrialized-country CO_2 emissions were down about 5 per cent from 1990 levels because of industrial collapse in eastern Europe. The Kyoto Protocol therefore represents a commitment for the industrialized countries to keep their aggregate emissions at 1995 levels. Combined with growth projections in the rest of the world, this means that the Protocol allows global greenhouse gas emissions to grow by more than 30 per cent. This is not an agreement that will get climate change under control, nor indeed one that is likely to have a huge impact on energy markets. Also, as indicated above, actual national emissions may be very different from the headline commitments.

The process of implementing the Protocol will be long and difficult. The first major political milestone after Kyoto was the next annual Conference of the Parties to the Climate Change Convention, COP-4, held at Buenos Aires in November 1998. In many respects COP-4 was an epilogue to the Kyoto Protocol itself, confirming the basic agreement and defining the follow-up process. It was widely referred to as 'dealing with Kyoto's unfinished business'.

In some ways the most significant developments occurred before COP-4 even convened, as it became clear that all negotiators accepted the Protocol (together with other issues under the convention) as the basis upon which they would work henceforth, and that the key agenda of COP-4 would be to set out a credible work programme. The specific result of the conference was the Buenos Aires 'Plan of Action', which consisted of decisions in six specific areas:[4]

- financial mechanism;
- development and transfer of technologies;
- addressing the specific needs and concerns of developing countries, including minimization of adverse impacts (Articles 4.8 and 4.9);
- activities implemented jointly under the pilot phase;
- the mechanisms of the Kyoto Protocol;
- preparations for the first session of the Conference of the Parties serving as the meeting of the Parties to the Kyoto Protocol.

Thus Buenos Aires succeeded in 'maintaining the momentum' of Kyoto, as many diplomats described it. In practice it did more than that: it added considerable clarity about the timescales and politics of the process, if little about the substance. Many important decisions are now focused upon the sixth Conference of the Parties, set for the year 2000 and likely to be held alongside the subsidiary body meetings in late October that year. COP-6 could indeed rival Kyoto itself in terms of the scale of the issues scheduled to be resolved there.

The realities of domestic implementation

Despite all the international discussion and expressions of concern and commitment, few countries have yet really succeeded in integrating climate change concerns beyond the fringes of energy policy and major energy investment decisions. Several OECD countries are unlikely to meet the targets in the Convention, and it remains unclear how far pressures developed from the international negotiations will be transmitted through national political systems. The reality is that many people involved in climate change have never understood the complexities and competing

[4] Decisions 2/CP.4 and 3/CP.4 on the financial mechanism, with others numbered sequentially in the order given here.

objectives associated with energy policy, and many of those involved in energy policy have never taken very seriously the objective of limiting CO_2 emissions. The wagon of international negotiations thus risks leaving the realities of implementation behind.

It would not be accurate to describe this as an impasse, or even a failure to introduce meaningful 'climate policies'. The problem lies partly in the fact that the issue arose against a background of 'environmental problems' in which it was possible to consider solutions separately from overall energy policy: specifying clean-up technologies or processes, or otherwise imposing requirements that led industry to bear the costs of implementing clean-up. Climate change is far more complex than this, because it is intimately bound up with overall energy policy.

The result is that 'climate change policies' considered to date cannot be driven and implemented by pressure just from a discrete climate change lobby or environment ministry. Rather, they become part of the broader process of coalition politics that dominates big decisions in any democracy. Policies that limit CO_2 emissions substantially are likely to be part of a package that attracts a range of other interests. Frequently, climate change has been a modest influence upon the outcome of decisions that had to be faced anyway. Most of the elements in the UK's national climate programme, for example, can be traced to coalitions that involve other interests, whether it is the tax-raising interests of the Treasury or the social concerns of the fuel-poverty lobby to see poor homes better insulated.

In most countries climate change is emerging as a diffuse influence upon other decisions, rather than a discrete policy area; an influence that pushes decisions in the direction of lower carbon and the promotion of measures to improve energy efficiency and clean energy technologies. The strength of this influence will be determined by popular perception, media concerns, the power and ability of environment ministries, and the extent to which the development of international law reinforces the governmental commitment to find ways of achieving targets or implementing particular kinds of policies.

So far, in most countries, the influence has been relatively weak, as illustrated by the 1995 emissions estimates set out in Table 3.1. Though the European Union announced in March 1996 that it expected to achieve the overall stabilization target for 2000 (though some further measures may be necessary),[5] and Japan expects to achieve its original per capita (but not absolute) target of stability at 1990 levels, the outlook in the United States is much less promising and several other countries (Canada, Australia, New Zealand and Norway) will clearly be wide of the mark. It is politically significant that, if the EU does achieve its goal, as is now claimed, this will mean that most industrialized countries will have attained the targets to which they signed up at Rio.[6] But critics can justly point out that most of the progress, even in those that do meet their targets, is attributable to factors other than climate policy – whether it is German reunification, economic transformation in eastern Europe or privatization of the UK electricity supply system.

Nevertheless, in many countries, the pressures and actions to limit CO_2 emissions are slowly gathering pace and volume. In my judgment, such an accumulation of small measures, perhaps growing into stronger measures as the institutional reach of climate concerns extends and the negotiations proceed, may have a substantial long-run impact. But for a long time, the politics of implementation, and the credibility with which countries can really deliver on promises they make, will come under increasing scrutiny as the international negotiations on implementing the Protocol proceed.

Conclusions

It is worth remembering that the Kyoto Protocol represents the culmination of twenty or thirty years of scientific developments internationally and ten years of political process. Most Western governments now have

[5] European Commission, *Second Report under the Monitoring Decision*, EC-DG-XI (Brussels, March 1996). The results of this assessment, announced to the Ad Hoc Group on the Berlin Mandate on 4 March 1996, projected on the basis of member states' submissions and Commission analysis that the total CO_2 emissions from the EU in 2000 would be in the range 0–5% above 1990 levels in the absence of a carbon tax, and stated that other kinds of additional measures could be brought forward to achieve the stabilization goal if it proved necessary.

[6] Fifteen countries are covered under the EU 'umbrella' target, and the Convention commitments are deliberately structured to allow the aim to be met jointly in this manner. The Cohesion countries (Greece, Spain, Portugal, Ireland), France and some others made it plain that they had only signed the Convention under this provision, and thus can claim compliance if the EU achieves the goal collectively. In addition, all the central and east European countries will have CO_2 emissions in 2000 below their 1990 levels. Strictly speaking, therefore, about 80% of Annex I countries will be able to claim compliance with the interim aim.

significant investment in the process and are convinced of the impotance of the problem. The Kyoto Protocol establishes the basic institutional structure and the first round of commitments. In my judgment that Protocol will eventually enter into force, although there will be a very difficult battle, particularly in the United States, to achieve this and it will not happen for three or four, maybe five years. A great deal of work will be done on how the flexible mechanisms are to operate, but in my view they will lead to market-based instruments for controlling emissions. Action at present may be quite modest, but will probably be strengthened over time and in pursuit of those objectives. Governments will be adopting quite a wide range of policies to meet their Kyoto commitments as part of the multiple goals of energy policy.

4 Implications of environmental initiatives for energy producers

Peter Kassler

The oil-exporting countries of the Gulf are, in financial terms, the world's most important energy exporters.[1] Even with limited exports from Iraq, in 1996 they achieved net energy exports worth some $115 billion.[2] The comparable figure for 1998 is estimated at about $70 billion, showing the enormous economic damage to Gulf countries resulting from that year's lower oil prices. Figure 4.1 illustrates the high level of dependency of the Gulf countries on energy exports. For comparison, the figure also shows the situation of Australia, which is a major coal and gas exporter but also has a large and diversified portfolio of export products and is therefore much less vulnerable to changes in energy demand.

The very high value of these exports is both a strength and a weakness, because, having attained a level of economic development through oil, the Gulf countries have become dependent on the fortunes of this one commodity market. They therefore feel themselves to be particularly vulnerable to the effects of policies designed to restrain oil demand in importing countries. The over-supply of the oil market in 1998 and early 1999 had much more to do with the Asian economic downturn than with today's environmental initiatives, but it is an example of what oil exporters fear may happen in future as these initiatives take hold.

[1] This paper is based on a research project carried out by Matthew Paterson and myself in the Energy and Environmental Programme at Chatham House, results of which were published at the end of 1997 in a book entitled *Energy Exporters and Climate Change: Policies, Impacts and Options*. The work was funded and sponsored by agencies of the Norwegian, Netherlands and Canadian governments and the United Nations Climate Change Secretariat.

[2] Throughout this paper energy reserve and production figures are derived from BP and BP AMOCO, *Statistical Review of Energy* 1997, 1998 and 1999. GDP and other economic figures are from *The Economist Pocket World in Figures* and *Pocket Middle East and North Africa* 1997, 1998 and 1999.

Figure 4.1: Gulf nations' energy and non-energy exports

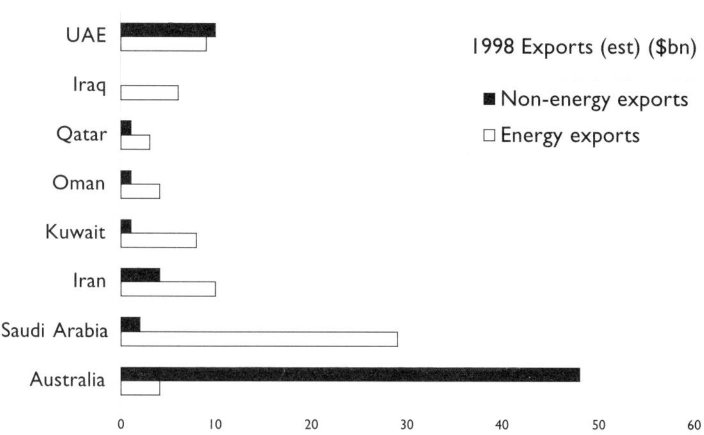

Source: Derived from *The Economist Pocket World in Figures,* 1999, and BP AMOCO, *Statistical Review of Energy,* 1999.

This paper therefore reviews:

- the nature of the Gulf oil exporters' concerns;
- the current political background to climate change initiatives, mainly in major consuming countries;
- some energy scenarios published by various bodies to illustrate the market impacts of environmental policies;
- current energy market developments, as a reality test of the scenarios;
- the vulnerabilities of the Gulf nations;
- what options oil exporters may have in order to sustain their economic welfare in the closing decades of the oil era, and beyond.

Gulf exporters' concerns

Since the earliest recognition outside specialized scientific circles that the combustion of oil, coal and gas might, through the production of greenhouse gases, be contributing to the warming of the global climate, the oil-producing countries have seen climate change policies as a threat to their well-being.

This perception stems mainly from their fear that the industrial countries might attempt to reduce their emissions of these gases by energy taxes and strict regulatory regimes controlling the use of fossil fuels in their economies, particularly for industry and transportation. Some of the exporters' main areas of concern about the economic impact of such initiatives on the upstream countries are as follows:

- reduced demand, price, revenue and GDP growth, compared with 'business as usual';
- reduced export revenues, increased import costs;
- loss of trade by oil producers to gas producers;
- replacement of fossil fuels by renewable energy;
- oil and gas reserves being left in the ground.

The Gulf countries' concerns are amplified by a number of other realities arising from the relatively recent arrival of their oil wealth. They all have ambitious growth and development plans requiring expenditure to be funded out of oil revenues. Their populations are young and fast-growing, the under-25s amounting to around half the population (see Figure 4.2).

Figure 4.2: Age profiles of the Gulf countries, 1995 (% of population)

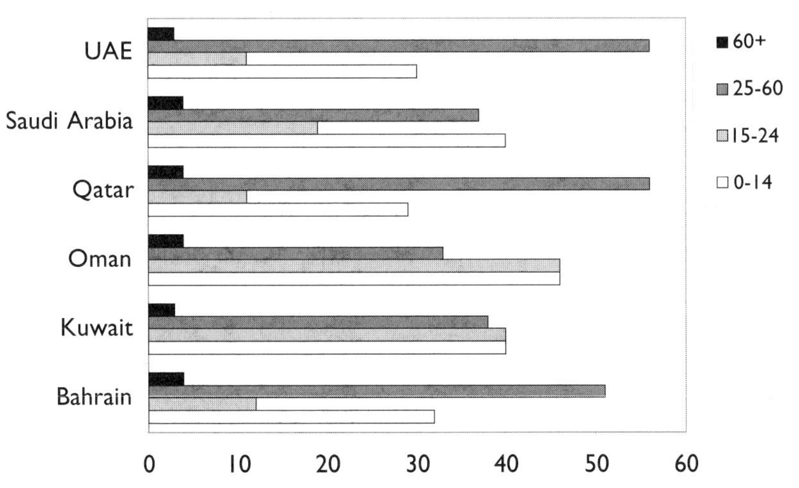

Source: The Economist Pocket World in Figures, 1997.

(For comparison, in a typical village in southern England, the population is roughly equally divided among under-25s, 25–60s and over-60s: on average, much older.) These young people are hungry for education and well-paid employment. And these nations feel themselves to be threatened by enemies outside, and, in some cases, within.

The Gulf is one of the world's most sensitive and politically volatile regions, exemplified by Iraq's conflicts with its neighbours, not very distant echoes of the Arab–Israeli struggle and wars in the Kurdish region and Afghanistan. These concerns lead Middle East states to spend a larger proportion of their wealth on armaments than most other parts of the world. Dr Mahmoud G. Salameh estimated recently that expenditure on defence by the Gulf countries over the period 1974–97 may have amounted to about one-third of their oil revenues.[3] For these reasons, any threat to the continuous future growth of their oil revenues leaves the Gulf oil exporters feeling particularly vulnerable.

Defence spending as a percentage of GDP is shown in Figure 4.3, which groups Middle East countries together with North African states.

Figure 4.3: Defence spending comparisons (% of GDP)

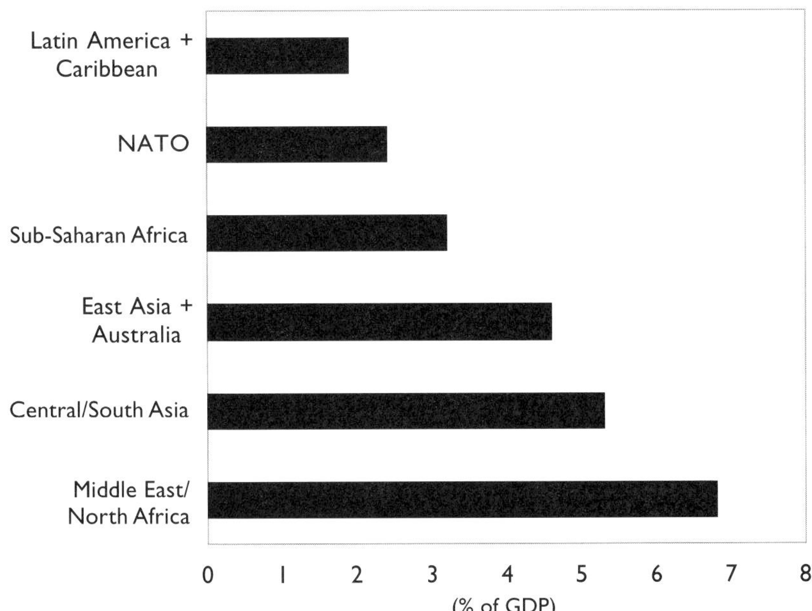

Source: Petroleum Review, August 1998.

Peter Kassler

Climate change politics

In Western countries, the 1990s have seen a gradual development of public and political consciousness about climate change. As described in Chapter 3, the major milestones have been the United Nations Framework Convention on Climate Change which emerged from the Rio 'Earth Summit' in 1992, the 1995 Berlin Mandate and the subsequent international negotiations, leading to the signature of the Kyoto Protocol at the end of 1997 and its confirmation at the Conference of the Parties in Buenos Aires in 1998. Political consciousness is not, however, the same thing as a commitment to action. The real issue now is how vigorous the responses to Kyoto are likely to be over the next few years, especially in the major oil-consuming countries.

Among the industrial nations, the largest importing region is North America, and here the political classes have remained sceptical about supporting domestic actions to limit greenhouse gas emissions. Their counterparts in Europe have generally accepted the desirability of some action, however reluctant they may be to undergo any personal loss as a result. Government actions here have often been ambiguous, as in the case of the UK which has continued each year to increase taxation on vehicle fuels, while also bringing in measures whose effect has been to subsidize coal production and restrict growth in the use of natural gas for power generation.

Since 1997 there has been a growing consciousness in the West of the economic problems experienced by a number of Asian and Latin American countries, accompanied by fears that these may spread to the rest of the world and trigger a global recession. Although lessened now by signs of recovery in Asia, this concern has tended to divert public and political attention away from the environment. There is also a sense that environmental commitments and measures in many countries 'belong' to environment ministries, and these may not be the most powerful agencies in their governments, which have to balance their concerns against other priorities including economic growth. As a whole, therefore, governments are reluctant to enact measures such as energy taxes and energy-use regulation which may have (or be accused by their opponents of having) a growth-reducing effect. In the United States this tendency has probably been reinforced by the inability of the Clinton–Gore administration, which had invested a good deal of political capital in promoting the Kyoto agreement, to convert this into real environmental measures.

This paper refers to several 'deep green' scenarios, including those published by Greenpeace, which project CO_2 reduction measures much more severe than any which are currently politically acceptable. The next section of this paper, on policy impact scenarios, discusses conditions in which these might become more realistic.

Throughout the pre-Kyoto negotiations the oil-exporting countries, led by a group of OPEC members, maintained a cool attitude to the whole process. Among the industrial countries a similar line was taken by Australia, a coal and gas exporter, and by major industrial interests in the United States, including some coal and oil companies, heavy manufacturing industries and the corresponding trade unions. It is interesting, though, that some major oil companies have detached themselves from the organized opposition to implementation of the Kyoto Protocol (Shell and BP both resigned from the Climate Change Coalition, an industry pressure group opposed to Kyoto) and are taking steps to manage greenhouse gas emissions through their own operations, including the setting-up of internal emissions trading systems.

To sum up these political reactions from the industrial world, it may be that Kyoto is an idea whose time has not quite come. The combination of fairly lukewarm public assent to environmental measures with the recognition of much more immediate economic concerns makes it unlikely that such measures will have severe short-term effects on energy exporters. But neither will they go away, and there is always a possibility, as discussed later in this paper, that the green movement may in future be revived by real climatic events. In the meantime, the uncertain price environment and low demand growth in the oil industry give exporters enough to worry about.

In the pre-Kyoto negotiations, the oil exporters initially opposed greenhouse gas limitation measures altogether, emphasizing the scientific uncertainties about climate change and their potential financial losses from these measures. During 1997 they changed the emphasis of their negotiating approach, which came to be based on a request for cash compensation from the industrial importing countries for any future losses of revenue resulting from oil not being exported as a result of the Kyoto Protocol (compared with a 'business-as-usual' demand projection assuming no Kyoto measures). This request was partly acknowledged in Article 3 of the Protocol, in which the oil exporters' situation is linked with that of other developing countries potentially vulnerable to climate change measures.

Our study of impacts and options was motivated by a feeling that financial compensation was not the best way to address the oil exporters' concerns, and in any case was demonstrated during the negotiations to be unacceptable to the importing countries. We also felt, however, that we should try to learn more about the scale and reality of the problem, and that there were perhaps other options available to address any real hardship experienced by the oil exporters. The importance of the Gulf to global security, referred to above, was also a factor in the international community's wish to resolve the concerns of these countries.

Policy impact scenarios

Long-term energy scenarios have been published by a number of institutions, including the International Energy Agency (IEA), World Energy Council (WEC), OPEC and Greenpeace. They are based on different sets of input assumptions and have been constructed for different purposes. The IEA, WEC and OPEC scenario sets contain pairs of scenarios roughly corresponding to 'business-as-usual' and environmentally driven worlds of the future, and the differences between them give some impression of the demand impact of 'green' policies, including greenhouse gas limitation measures between now and the middle of the next century.

Table 4.1 shows the range of long-term projections of oil demand derived from the published scenarios. Although these scenarios incorporate a wide range of assumptions, for this purpose they have been grouped by

Table 4.1: Oil market scenarios

Scenario	Oil demand (1990=100)	
	2010	2020
Business-as-usual	135	145
Moderate Green	125	130
Deep Green	70	55

outcomes into those incorporating moderate application of CO_2 limitation measures ('Moderate Green') and those invoking much more severe application ('Deep Green').

The severe environmental case, in which demand for oil is halved in 30 years, would be very bad news for the Gulf countries. This could easily lead to prolonged production over-capacity in major producing countries, and price collapse. In turn, this 'world' would imply a need to plan for radical restructuring of economies which have become very dependent on oil. We will return shortly to the conditions in which such scenarios might occur.

In the 'moderate green' worlds of these scenarios the issue will be lower growth of demand than in the more optimistic 'business-as-usual' projections. In the cases cited above, the annual growth in demand would be only 0.5 per cent per year less than in the full-growth cases. These look quite survivable for the oil exporters, requiring much less radical adaptive planning, mainly limited to appropriate timing of projects to increase oil production capacity.

Projections suggest that in such moderate scenarios, most of the world's resources of conventional oil would be consumed, avoiding leaving large unused reserves in the ground. This reflects estimates of the duration of the current competitive advantage enjoyed by oil-based fuels, mainly for transport applications.

Outcomes for other fuels according to the various scenarios may be summarized as follows:

- gas demand grows in most scenarios, with slight gains of market share in moderate and deep green projections;
- long-term prospects for coal are poor in both green cases;
- renewables demand creeps up very very slowly.

We found it surprising, reviewing these scenarios, that they did not show more relative growth in gas demand as a result of global warming concerns, and do not know whether this reflects the reality of very slow change of global energy use patterns, or a lack of imagination. For a gas-exporting country such as Abu Dhabi, it is important to take a view on this subject.

The real world of energy

If energy demand were not influenced by politics, the weather, human fashions, economics, wars, resource discoveries and technology, among a number of factors, it would be easy to forecast. It is; and it is not. Nevertheless, over a 20–30-year period the big unknown variables are probably recovery of economic growth, particularly in Asia, and the influence of the environmental movement, including its impact on everyday transportation and domestic technologies.

Figure 4.4 shows regional changes in energy demand from 1997 to 1998. On this time-scale, the business cycle is the most important factor in determining changes in energy use, although the figure also hints at some longer-term features. World primary energy and oil demand have both grown at more than 1 per cent per year for the past ten years, but in 1998 both were stagnant, reflecting the economic problems in Asia and the Russian financial collapse. There has, however, been a stable trend for gas to gain market share, mainly at the expense of coal and to a lesser

Figure 4.4: Energy demand change, 1997 to 1998 (%)

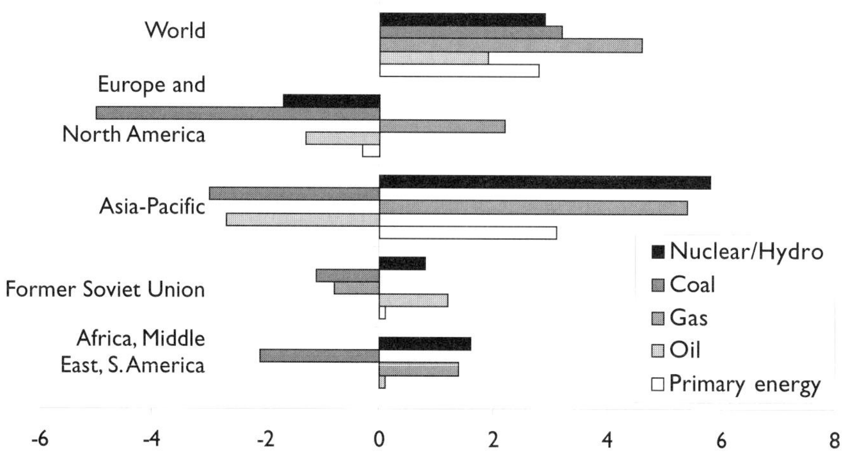

Source: BP AMOCO, *Statistical Review of Energy*, June 1999.

degree of oil, largely for environmental reasons. This was maintained in 1998 although the actual growth of gas demand was also below its long-term trend of about 2 per cent per year, for the same cyclical reasons as those for oil.

The regional breakdown of demand changes confirms that the only regions with significant growth in oil demand were Europe and North America and the much smaller markets of South America, Africa and the Middle East. This is very different from the picture of most of the last decade, in which these figures were dominated by growth in Asia.

A period of more than a year in which oil demand has slumped has been very difficult to manage for the OPEC member states. Their market management and production quota allocation has been predicated on a previously reasonably reliable annual growth in world demand for their product of nearly 1 million barrels per day, which in turn has been the driving force behind their internal budget planning and expenditure decisions. The combined results of low prices and lack of growth in sales led in 1998 to a decline in the Gulf exporters' oil revenues of about 40 per cent compared with normal levels, resulting in severe difficulties in maintaining the expected levels of expenditure on defence and social provision.

These difficulties, as already indicated, were not caused by energy taxes or other environmental measures, but they provide an example of

Figure 4.5: Industrial countries' CO_2 emissions, 1990 and projections to 2000 (million tonnes)

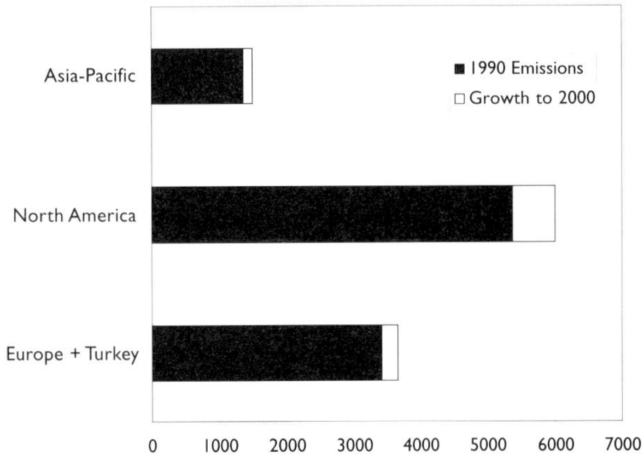

Source: IEA.

what life would be like for oil exporters in a world in which demand was severely restricted by measures of this type. However, this has not yet proved to be the case in the major consuming countries. At Rio in 1992 the industrialized countries ('Annex I signatories') engaged themselves to promote measures to restrain their CO_2 emissions in 2000 to 1990 levels. Figure 4.5, based on projections by the IEA, shows their current performance against this target.

The figure shows that CO_2 emissions from this group of countries are still growing, and that the 2000 total is likely to be 10 per cent above that of 1990. This suggests that there is still a sizeable gap between the political rhetoric about greenhouse gases and the reality in energy markets, and that the impact of environmental policy on energy consumption is likely to be slow and gradual.

In its 1998 analysis, the IEA pointed out that:

> There are many possible combinations of energy saving and fuel substitution that meet the Kyoto commitments. But all involve deviations from past trends. It is clear that they will not happen unless adequate policies and measures are put in place by governments to make them happen. These policies will involve considerable practical difficulties, not least because of the relatively short time remaining to the period 2008–2012 in which the Kyoto commitments are to be met.

It would nevertheless be unwise totally to write off the possibility of a 'discontinuity' causing a collapse in the demand for fossil fuels. Various events, technological, political or climatic in nature, could combine to cause this to happen, for example:

- a succession of climatic or ecological disasters, finally persuading the public that global warming is real;
- political leadership which credibly promotes environmental policies;
- technology-promoted fuel economy through commercially viable fuel cells, etc.;
- economic recession in major consuming countries.

If several or all of these things happened together, the effect, particularly on oil demand, could be quite powerful. Events in the United States will be particularly important. Here, there are two factors which could come into play:

- the huge potential for energy conservation represented by the transport sector;
- the fact that oil conservation would also contribute to national security by reducing the currently growing dependence on the Middle East.

This is illustrated by Figures 4.6 and 4.7.

Figure 4.6: US dependence on oil imports (Mbd)

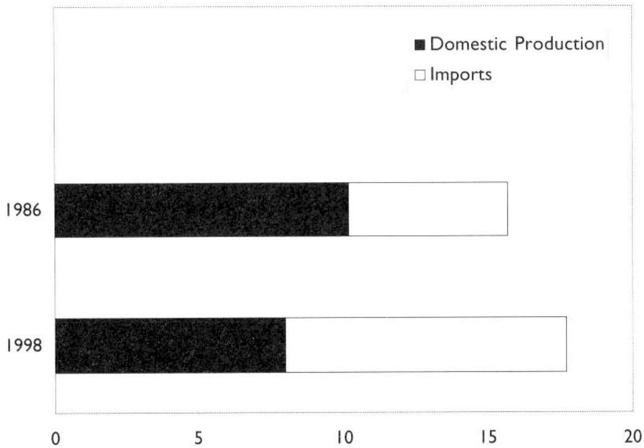

Source: BP AMOCO, *Statistical Review of Energy*, June 1999.

Figure 4.7: Consumer prices for gasoline, 2nd quarter 1997

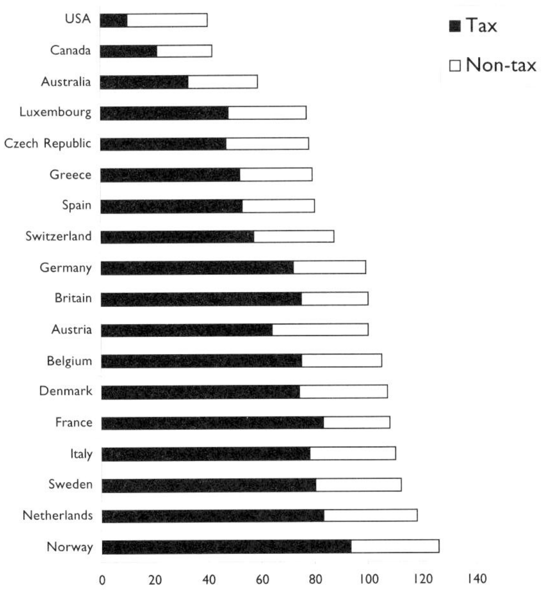

Source: IEA

Figure 4.6 shows that considerably more than half the United States' oil consumption now originates as imports. In spite of the great technical successes achieved in the deep-water Gulf of Mexico, this trend seems set to continue inexorably, unless something happens to change the supply–demand balance.

Figure 4.7 shows that the United States and Canada enjoy extremely low gasoline prices, mainly as a result of low taxes on this fuel. It illustrates just how little incentive there is to economize on petrol consumption there, and largely explains the size and power of the vehicles in daily use. The United States and Canada are currently the centre of a large number of technological initiatives aimed at the production of the twenty-first-century 'supercar' with very low fuel consumption (see Chapter 2). There are indications that some of these are successful and attracting commercial investments from major motor manufacturers and even oil companies.

The United States is in a unique position in the oil market, as a very large oil user, with both the reason (security) and the means to reduce its consumption. It is quite credible that it will do so one day, and that the results will have a major impact on the global market for oil.

One conclusion to be drawn from these reflections on 'the real world of energy' is that, in the short term, events in global energy markets largely reflect the ups and downs of the business cycle. There are, however, underlying political, economic and technological developments in progress which, in the longer term, may lead to profound restructuring of the world's energy economy. The result could be an overall gradual change in energy use, replacement of existing technologies by others consuming less fossil fuel or even based on renewables, and hence low energy prices. This would have very serious implications for the Gulf nations.

Vulnerabilities of the Gulf nations

Figure 4.8 shows the generally high proportion of national wealth in the Gulf countries which is generated by energy exports (25–40 per cent), and the very wide range of per capita incomes which they enjoy.

Figure 4.8 is mostly based on 1996 national income statistics, and the GDPs of oil-producing countries were certainly lower in 1998, but the point is still valid. Australia and Canada have been included in the chart

Figure 4.8: Energy export dependency and national wealth

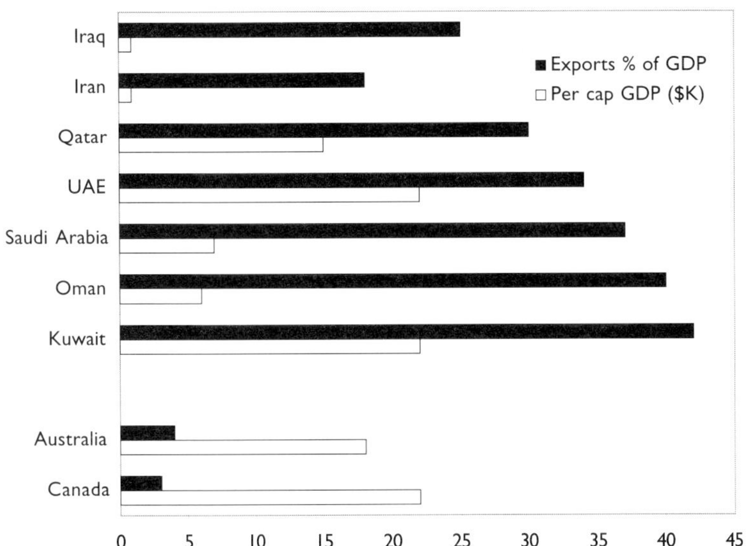

Source: The Economist Pocket World in Figures, 1997/8 and BP AMOCO, *Statistical Review of Energy,* 1998.

to show the very different profiles of affluent industrial countries which are also energy exporters; because these have much more diversified economies than the Gulf states they are much less dependent on the export of their coal, oil and gas.

Clearly Iraq and Iran, with per capita GDP below $1,000, are particularly vulnerable to any significant decline of their oil revenues, but even the richer Gulf countries have precariously balanced national finances in which spending plans are governed by the oil price assumption in their national income forecasts.

The view in the Gulf monarchies is believed to be that much government expenditure is necessary to preserve national cohesion; it is apparently perceived as one of the ways to prevent radical disaffection within the population. It is therefore very difficult to change the amount and pattern of this expenditure. In the longer term, this is one of the assumptions requiring dispassionate review.

Options for Gulf exporters and importing countries

In the climate change negotiations, Gulf states asked to be compensated in cash for any decrease in oil offtake attributable to the Kyoto agreement, compared with a 'business-as-usual' projection. Apart from the debatable logic underlying this calculation and the technical difficulty of negotiating and performing it, compensation is unlikely to be adopted because the sums of money potentially involved are very high (of the order of the total of current OECD civil aid budgets to developing countries), and the recipients have less popular appeal and political resonance in Western countries than the poor nations currently being helped. There is, nevertheless, a feeling among some Western governments that ways should be found to help the oil producers, if indeed they can be shown to have suffered as a result of Kyoto. These feelings are rooted partly in feelings of equity, and partly in the view that it is important to include the energy-rich, politically volatile and sensitive Middle East in an international consensus about energy management and the environment. This view led to a quest for alternative negotiating agendas between exporters and importers.

These options reflect a search for ways in which the Gulf countries might wish gradually to adapt to a world which will slowly have less need of their main product (although along the way there may well be a tendency for the less prolific non-OPEC fields to start to run down, and demand for the Gulf production will still be very high).

Another theme is the recognition that the Gulf states make some very fair points in the negotiations about the way in which Western countries tax and use fossil fuels.

The following are some of the options which might appear on such an agenda.

Taxation and subsidies to reflect carbon content of fuels

If the 'green' ideal is an energy economy in which taxation reflects the carbon content of the fuels used, the global reality is far from that. Most industrial countries (except the United States and Canada) load taxation on oil, because it is easy to collect. Coal, a polluting high-emitting fuel, is often subsidized. This imbalance could be rectified by reform of energy taxation in consuming countries.

Well-head taxation

Exporting countries would love to impose upstream environmental taxes and use the proceeds (the 'well-head tax'), but have not worked out how to do so since such a tax would, in effect, be an increase in the market price of oil and would expose the tax-levying country to competition from other countries without a well-head tax, as well as from competing fuels. This would be particularly difficult in current circumstances owing to the global over-supply of oil.

International emissions trading and quotas

A system of international emissions quotas and trading would allow global CO_2 reduction efforts to be focused on the most cost-effective solutions (such as cleaning up dirty industries in former Soviet-bloc countries). If the Gulf states were included this would permit a number of initiatives, including the 'greening' of oilfield operations, as has been done in Norway, and the sale of 'green oil', linked to CO_2-absorbing forests grown in projects financed or sponsored by OPEC countries. Surveys in some Western countries suggest that a substantial proportion of consumers are prepared to pay a premium for energy products with 'green' credentials.

Gas exports to replace oil and coal

An increase in gas exports to supplement and partly replace oil and coal for power generation and transport uses would benefit those oil-exporting countries that also have gas resources and would stimulate exploration for gas. This is probably the most attractive 'large' option from an environmental viewpoint.

Alternative energy developments by petroleum exporters, including solar systems

A number of exporting countries have climates with enough sunshine to support solar thermal power generation for domestic use or export.

CO_2 disposal

Oil-exporting countries have the sort of large, simple geological structures which would be suitable for CO_2 disposal. The CO_2 could be collected from some existing industrial and chemical processes. Upstream separation of CO_2 also increases the attraction of the future 'hydrogen economy', perhaps using advanced fuel-cell vehicles powered by hydrogen or methanol made from hydrocarbon feedstocks.

Diversification of exporting economies

Finally, oil-exporting countries may wish to consider diversification of their economies, finding new sources of competitive advantage. One factor in this shift could be the possession of hydrocarbon stocks from which high-value non-fuel products could be manufactured.

None of these options calls for large-scale subsidies from consuming governments, although some of the 'green' technologies may require seed money through the pre-commercial stages. Rather, they present options and opportunities for investment, which may, if the political climate is right, be seized by private firms with appropriate technologies and personnel. They also present a major challenge to regenerate networks of international cooperation that were placed under strain by the very different interests and priorities which surfaced during the climate change negotiations.

5 Environmental developments in the United Arab Emirates: achievements and challenges

Saif M. Al-Ghais

There has been unprecedented global economic growth in the second half of the twentieth century. As the economy has grown, so pressures on the earth's natural systems and resources have intensified, resulting in sharp increases in the consumption of food grains, seafood, water, forest products, fossil fuels and many other natural resources. The unfortunate reality is that a continuously expanding global economy is slowly destroying its host – the earth's ecosystem. An environmentalist has described this growing relationship in terms of the metaphor of a growing cancer cell, which eventually destroys its own life-support systems by destroying its host. Ecosystems worldwide continue to be under unrelenting stress caused by deforestation, aquifer (water table) depletion, habitat destruction, over-fishing, urban development, temperature rise, food scarcity, industrialization, agricultural growth, disappearance of wildlife and environmental pollution. The coupling of population control with sensible strategies of environmental and ecological management will be crucial to achieving a sustainable prosperity for humans, while still accommodating other species and their natural communities.

The UAE: historical and geographical background

Under the leadership of His Highness Sheikh Zayed Bin Sultan Al Nahayan the United Arab Emirates was created in 1971 as a federation of seven Emirates: Abu Dhabi, Dubai, Sharjah, Ras Al Khaimah, Ajman,

Fujairah and Umm Al Qaywain. All the Emirates are located on the eastern edge of the Arabian Peninsula and together they cover a total area of about 83,600 sq. km. The population of the UAE has increased from about 180,000 in 1969 to about 2.5 million in 1997. The total length of the UAE coastline bordering the Arabian Gulf area is approximately 650 km. Geographically, about 65 per cent of the country may be classified as sandy desert, 14 per cent as inland *sabkha* (salt marsh), 13 per cent sand mixed with gravel, 3 per cent coastal *sabkha* and 5 per cent largely stony or planted area. The maximum and minimum recorded temperatures are 47–48°C and 9–10°C, respectively, with high humidity occurring along the coast throughout the year, reaching a peak in July and August. The rainfall over much of the country is scant (less than 100 mm annually).

In the four decades following the discovery of oil in the 1950s there have been dramatic and enormous transformations in all spheres of life, in terms of both social and material advances. During this period there has been rapid urbanization, coupled with the development of modern and prosperous living conditions, and marked progress in agriculture and industry, especially in the oil sector.[1] The Emirates have many terrestrial and coastal habitats that are important locally, regionally and globally. As is only to be expected, urbanization and economic growth are likely to affect the local ecosystem and wildlife.

Increasing awareness in the UAE of the need for environmental protection, conservation and management of natural resources, and the maintenance of biodiversity stimulated the initiation of a process aiming to reform national policies, plan various environmental development programmes and create governmental and non-governmental organizations, agencies, and research and teaching institutions dealing with environmental issues of national interest. The remainder of this paper discusses some examples of environmental developments, activities and operational areas.

[1] P. Vine and P. Casey, *United Arab Emirates: Heritage and Modern Development* (London: Immel Publishing, 1992).

Plantation

The arid climate, nutrient deficiency, low rainfall and extreme summer temperatures are hostile to plant growth in this region. Besides date-palm trees, around nine or ten local plant types representing species of *Cyperus, Haloxylon*, and *Acacia* were found to dominate the desert vegetation. Some of these species support traditional grazing and medicinal practices.[2] In the towns, however, as a result of the plantation initiatives launched by the municipalities and supported by the water-dripping irrigation system, there has been extensive greening of urban areas, which now support a wide variety of arid and tropical plants. In Abu Dhabi alone, for example, the date-palm tree coverage increased from 5 hectares in 1975 to 2,310 hectares in 1995. Now the dense bushy vegetation along roadsides and attractive gardens carpeted with grass are features of UAE towns that also provide habitats for native wildlife and migrant species. The municipalities, in collaboration with the Ministry of Agriculture and Fisheries, organize an environment and plantation awareness week every year, in which residents also participate.

Wildlife protection

The coastal areas support a range of wildlife including seabirds and migrant waders, sea-turtles and dugongs, as well as commercial fishery. However, terrestrial habitat degradation and undisciplined killing over the last 40 years has caused the decline or eradication of over 50 species of rich and diverse land mammalian fauna, including the Arabian leopard, wolf, Arabian gazelle, Arabian oryx, Arabian tahr and some species of poorly studied carnivores, cat and fox.[3] The country is internationally important for its bird life, located as it is on the west Asian migratory route for many species, including shorebirds and other waterfowl,

[2] R. A. Western, *The Flora of the United Arab Emirates: An Introduction* (Al Ain: UAE University, 1989); P. E. Osborne, ed., *Desert Ecology of Abu Dhabi* (Oxford: Information Press, 1996).

[3] D. L. Harrison and P. J. J. Bates, *The Mammals of Arabia*, 2nd edn (Sevenoaks: Harrison Zoological Museum, 1991); P.E. Osborne, ed., *Desert Ecology of Abu Dhabi* (Oxford: Information Press, 1996).

passerines (e.g. falcon) and birds of prey (e.g. houbara). Over 400 species of birds have been recorded in the UAE. Increasing pressure of urbanization leading to habitat degradation, combined with hunting of the bird population, especially houbara and falcon by traditional Arab falconry, and illegal trapping, have resulted in declining populations both in the UAE and in neighbouring countries. Terrestrial reptiles contribute significantly to the biological diversity of the UAE, being represented by at least three dozen species, of which *Acanthodactylus* and *Bunopus sp.* are the most common lizards and *Lytorhynchus* and *Eryx sp.* are the commonest and most widespread snakes. The locally occurring reptiles are morphologically and physiologically adapted to the harsh desert environment. For example, to enhance locomotion on soft sand they have smooth skins, pronounced pectination on digits and narrow, pointed heads.

Knowledge of the invertebrates has been comparatively poor in the Middle East. Recent studies by the National Avian Research Centre (NARC) of the Environmental Research and Wildlife Development Agency (ERWDA) have shown that the most common and widespread taxa occurring in the desert ecosystem and forming the greatest biomass are from the *Arachnida* and *Insecta* classes of arthropods, which are represented by over 200 species of desert beetles, mites and ticks, true insects, scorpions, damselflies and dragonflies, grasshoppers and bush crickets. These invertebrates are important in the food chains of birds, reptiles and mammals.

Wildlife reserves have been established at Sir Bani Yas, Qarnen and other islands and sites of the UAE, in order to preserve biodiversity and protect wildlife including marine turtles, dugongs, oryx and other terrestrial animals, and avian populations including seabirds. Under the instructions of Sheikh Zayed wildlife restoration programmes are being carried out for oryx, houbara, falcon and sea-turtles.

Marine resources

The clustered islands and shallow bays along the UAE coast provide a rich marine habitat such as mangrove forests, seagrass beds, algal mats, *sabkha*, *khor* (harbour) and coral areas.

The total world fish harvest, including freshwater and aquaculture, increased from 21 million tons in 1950 to about 116 million tons in 1996, which also included 23 million tons from aquaculture.[4] However, the total catch from the ocean has been declining, since attaining a plateau at about 90 million tons in the early 1990s. Most marine fisheries around the world are in danger of severe depletion. Total fish production in the UAE is about 90,000 tons per year. In the recent past, the marine fisheries resources of the UAE coastal waters have been increasingly exploited for commercial purposes. In addition, growing marine pollution and deterioration of mangrove forests, sea-grass beds, coral reefs and other marine habitats, which serve as nursery grounds for many crustaceans and fish, have started to cause problems for marine fisheries in the UAE, which are catching decreasing amounts of fish despite a rise in the number of boats and fishermen. To manage and develop fisheries in a sustainable manner, various measures and strategies have to be devised. These will include a government ban on fishing with trawlers; reduction of post-harvest losses through improved handling, processing, transport, distribution and storage systems; protection of marine ecosystems from deterioration; and systemic study, research and planning for fish stock assessment and improved fishing techniques.

Furthermore, the studies undertaken at the UAE University and ERWDA on the biology and habitat of local fish populations have led to their characterization and management.[5] The first comprehensive and systemic study of the biology, ecology, distribution and migration of sea-turtles is in progress at ERWDA, under the supervision of the author, aimed at developing measures for conservation of these fascinating but endangered creatures.[6] In recognition of the importance of the mangrove ecosystem for the development of fishery and other resources, the author has undertaken extensive studies to understand the physical, chemical and biological characteristics of mangrove (black or grey mangrove) swamps, their distribution and their importance in maintaining the local marine ecosystem. The crustacea of the mangrove ecosystem have also been

[4] FAO, *Fishery Statistics: Catches and Landings (1995),* FAO Fisheries Series No. 48 and FAO Statistics Series, Vol. 80, No. 134 (Rome: Food and Agriculture Organization of the United Nations, 1997), p. 713.

[5] S. M. Al-Ghais, 'Some Aspects of the Biology of *Siganus canaliculatus* in the Southern Arabian Gulf', *Bulletin of Marine Science*, 52 (1993), pp. 886–97.

[6] ERWDA, 'Turtle Conservation is on the Move', *Arabia Underwater World*, No. 1, 1997.

studied.[7] ERWDA has recently launched a mangrove plantation programme to restore and expand this vegetation in the Emirate of Abu Dhabi. Efforts are under way to introduce a new, large-sized variety of mangrove (red mangrove) from Southeast Asia to the Gulf region. Non-governmental organizations such as local business houses have also been contributing to these environmental development activities. Mangroves and turtle conservation projects were partly funded by Shell Abu Dhabi and Shell Middle East Marketing. ERWDA is also working with the national oil company, ADNOC, to develop protected beaches in certain areas of Abu Dhabi affected by oil operations.

Water resources

One of the most important basic human needs is an adequate supply of safe and clean water. However, a spreading scarcity of water is one of the major under-rated issues facing the world as it enters the third millennium. Growing consumption of ground water has led to the lowering of water tables on every continent, including in the Middle East. Because of increasing demand for water and low rainfall, there has been a continuous decline in the underground water table in the UAE. Realizing the seriousness of this problem, the UAE government has developed a number of water reservoirs, particularly in the northern part of the country, where rain water is collected and later distributed to dry areas for crop irrigation. Remarkable progress has been made in the development of water resources and management in the UAE, where adequate supplies of hygienic water are available from the sea (using desalination plants) and from ground sources for drinking, irrigation and other purposes. In the UAE the bulk of water for irrigation is supplied by treated sewage effluent.

[7] S. M. Al-Ghais, *The Mangrove Ecosystem in the United Arab Emirates*, a report submitted to Shell Gas Abu Dhabi BV Shell Markets Middle East, UAE (1996); S. M. Al-Ghais and R. T. Cooper, 'Brachyura (Grapsidae, Ocypodidae, Portunidae, Xanthidae and Leucosiidae) of Umm Al-Quwain Mangal, United Arab Emirates', *Tropical Zoology*, 9 (1996), pp. 409–30.

Environmental research and literacy: the role of government and NGOs

In order to support environmental planning, development and management, the federal and local governments of the UAE have established environmental agencies and institutions, framed environmental laws and ordinances, and encouraged public awareness programmes.

The Federal Environment Agency (FEA) was established in 1993 to develop a clean and healthy natural environment and determine the necessary plans and policies to safeguard it from the damaging effects of pollution on human health, agricultural crops, wildlife, marine life and other natural resources. The Environmental Research and Wildlife Development Agency was established in 1996 by the government of Abu Dhabi to protect the environment, natural life and its biological diversity. The National Avian Research Centre, which was established in 1989 in Abu Dhabi as a scientific organization for the conservation and propagation of bird populations and associated ecosystems, is now a part of ERWDA. The multidisciplinary studies undertaken by the centre have contributed considerably to the knowledge of biology and ecology of houbara, falcon and other local birds, their health management and breeding conditions, which are monitored by closed-circuit television systems. The ERWDA has been using Geographical Information Systems (GIS), a sophisticated computerized database system, to study and store information about the ecology and habitat of birds and other wildlife, including satellite tracking of houbara and mapping of topography, human activity and the natural habitat of the UAE. To monitor the migration patterns of houbara and falcon between the UAE and the Central Asian states of Kazakhstan, Turkmenistan, Kyrgyzstan, China and Iran, the most advanced technique of satellite tracking of individual birds fitted with transmitters was successfully employed for the first time in a collaborative international research programme.[8] Lately, this technique has also been employed to track the movement of marine turtles in the Arabian Gulf. The study of locally occurring vegetation and of wild

[8] F. Launay, O. Combreau and M. Al Bowardi, 'Annual Migration of Houbara Bustards (*Chlamydotis undulata macqueenii*) from the United Arab Emirates', *Birds Conservation International*, 9, 2 (1999), pp. 155–61.

animals (mammals, reptiles and invertebrates) has provided comprehensive information about the flora and fauna of this region and their conservation. The marine environment section of ERWDA is working on the biology and ecology of fish, sea-turtle and other marine organisms, the conservation of mangrove forests and other marine flora, and the assessment of inorganic and organic pollution in marine abiota and biota and its deleterious effect at the cellular and molecular levels.

The local municipalities have overall responsibility for the development and management of the sewerage systems of the UAE in such a way as to maintain a clean and healthy environment. In addition, the Food and Environmental Control Centres of the municipalities are well developed to monitor the quality of edible products under existing consumer protection programmes. Checkpoints at the entry ports keep a strict watch on the quality of food products entering the country. Other activities supported by the municipalities include environment programmes for air, water and soil pollution, Environment Week (15–20 March), and the annual plantation drive conducted with the cooperation of the Ministry of Agriculture and Fishery and residents of the area. Some other environmental occasions celebrated annually at the national level are National Environment Day on 2 February, Regional Environmental Day on 24 April, World Environment Day on 5 June and Arab Environment Day on 14 October.

Non-governmental organizations have played an important role in protection of the environment and conservation of wildlife and natural resources. The Emirates Natural History Group, which was founded in 1976, has contributed significantly to this work by undertaking several studies and publishing a monthly magazine, *Tribulus*.

Sewerage system and treatment facilities

Organic waste is no longer seen as disposable garbage but as a soil-building natural resource. Re-use of organic matter can supplement fertilizer use while reducing the waste disposal needs of urban areas. Discharge of sewage into the sea without treatment, common in many tropical and subtropical countries, creates a public health risk through consumption of seafood and bathing.

The major urban centres of Abu Dhabi, Dubai, Sharjah and Al Ain, which accommodate nearly 80 per cent of the total UAE population, have well-developed sewer networks and sewage treatment facilities. The construction of the first major waterborne sewerage scheme in the country was commenced in the mid-1960s in Abu Dhabi. Almost all treated sewerage effluent produced is used for plant irrigation. According to a rough estimate, the sewerage treatment plant capacity in Abu Dhabi increased from 104,000 cu. metres per day in 1982 to 260,000 cu. metres per day in 1995.

Recycling of waste products

Recycling of waste products is an effective method of contributing to the conservation of natural resources and making the environment pollution-free, clean and healthy. This transition from the 'throwaway economy' to a re-use/recycle economy has made good progress in the United States and a number of European countries, where scrap metal, paper, glass and other wastes are recycled to meet the growing needs for these products. Recycling, too, creates new jobs. In fact, the use of renewable natural resources has been practised by Emirates people in their day-to-day life and habitat for a long time: for example, palm trees, stems and corals have been used for house construction in the past. Recycling plants for steel, aluminum, paper, plastic, and industrial and household waste products have been set up in the major towns of the UAE by private firms. ADNOC has set up a plant for separation of organic waste in the UAE.

Air, water and soil pollution

Both government organizations, such as FEA and ERWDA, and municipalities and NGOs participate actively in environmental monitoring and impact assessment programmes to tackle air, water and soil pollution. With increasing urbanization, population growth, industrialization (particularly the oil industry) and rising numbers of automobiles and other anthropogenic sources of emissions, the levels of hazardous air pollutants are increasing. These include carbon monoxide, nitrous oxides, sulphur dioxide, hydrogen sulphide, ozone, hydrocarbons, chlorofluorocarbons, lead and particulates. An increasing use of pesticides in agriculture, gardens and forests is also a threat to human and animal health. Attempts are in progress to create public awareness of these issues through the media, exhibitions/workshops and teaching/training programmes. The aim is to gain public support for measures to prevent pollution, such as reduced use of fossil fuels (e.g. coal and oil), increased use of non-polluting energy resources (e.g. solar and wind energy), the development and use of mass transit systems, anti-smoking drives, use of energy-efficient vehicles using exhaust treatment filters and high-quality lead-free petrol, and the setting up of waste product treatment plants in industries and plantations.

Efforts are also being made to improve natural resources and the productivity of soil, and protect it against pollution and deterioration. Various programmes undertaken to achieve this goal include restrictions on the felling of trees (a punishable act under local laws), construction of proper irrigation facilities, the planting of green belts of bushy trees surrounding agricultural and residential areas, regulated use of pesticides, planting crops beneath trees (agroforestry) and conservation of the land ecosystem.

Major challenges

In summary, the major challenges to the UAE's marine environment and its natural resources are activities associated with sea transport, particularly shipping and oil transport; eutrophication arising from over-enrichment of sea water with organic waste; coastal development; global warming; and pollution. The terrestrial environment is threatened mainly by the lowering of the underground water table, urbanization and other anthropogenic activities, population growth, pollution and global warming. These challenges have been recognized by the UAE government and its agencies, and programmes and policies have been undertaken in order to protect the UAE marine and terrestial environments.

6 Water shortages and conservation in the Gulf region

Roger Cragg and Hywel Thomas

The need for all nations to move towards developing sustainable development policies, particularly with regard to carbon dioxide emissions, has been emphasized by several contributors to this volume. Of equal importance is the sustainable use of water. It was suggested at a special session of the United Nations General Assembly held in June 1997 on Agenda 21 ('Earth Summit + 5') that if the present schemes for development and use of water resources were to continue, almost two-thirds of humanity might suffer from a moderate to serious water shortage before 2005, compared to one-third at present.

This paper will examine the definition of 'water shortage' and its two components, namely the available resource and the demand for water. The conservation of water, that is, ensuring that every drop is used wisely, is then considered, together with the options available to control demand when it exceeds the resources.

With limited natural water resources and spiralling water demand, the Gulf region already suffers a serious water shortage. The deficiency is made up with desalination, creating an apparent abundance. Is this sustainable in the long term?

Water shortage

The term 'water shortage' can have two meanings. The first is a short-term shortage caused by a failure in the supply system or, in the slightly longer term, a lack of water abstraction schemes to meet increasing demand. The other meaning is when demand is greater than the total available resources of a region. It is this second use of the term that we consider further here.

Water resource

What is the total water resource? Defining the extent of an available water resource is a prerequisite to being able to manage it effectively. Our understanding of weather systems and aquifers is improving all the time. However, when looking at the world's resources, the lack of reliable data soon makes it clear that there is need for an international integrated water information system. This is necessary in order to provide the information required to manage the water resource effectively. The information collected needs to meet internationally agreed standards and be presented in a format allowing easy and free exchange of data. Of particular relevance to the Gulf region would be the long-term monitoring of aquifers. As aquifers and rivers do not respect political boundaries, all countries which have an interest in a water resource will benefit from consistent and high-quality data.

As ground water is the main water resource in the Gulf region, it is vital to understand the effects that abstraction and discharges are having on the aquifers. The governments of the area should support international scientific and technical cooperation to develop common research methodologies for use in understanding the water cycle. Determining the safe and sustainable yield from an aquifer and where best to abstract is clearly vital to this region. The demise during the twentieth century of well-known artesian wells of Bahrain, with a recorded history going back for centuries, provides a frightening demonstration of the relatively rapid effects of the misuse of aquifers.

Rain water is a further available resource but, with limited precipitation of only around 80–100 mm per annum, the infrastructure needed to collect and store much more of it is unlikely to be feasible. Only the rainfall that does not already go towards recharging aquifers is available as a resource. Dams have been built across wadis to encourage the rain water to infiltrate into the ground and recharge the aquifers. However, a fair proportion of the water evaporates before it can enter the ground.

The remaining 'traditional' source of water in this region is desalination of sea water or brackish ground water. While in theory this is a limitless supply, the cost of production and the effect on the environment of CO_2 emissions from the fuel used to power the process mean that in effect this source is really only viable for providing high-quality domestic and industrial water. The need to control CO_2 emissions and reduce energy costs is leading to research into the development of new competing sub-technologies (e.g. novel membrane systems such as membrane distillation, and electro-chemical ion exchange). Links between existing desalination technologies and both new and proven renewable energy technologies (e.g. solar and wind) are also being investigated.

Alternative sources of power may improve the situation. Little research appears to have been done on the viability of wind power in the Middle East. The obvious alternative for the Gulf has to be solar power, given the large areas of land available and the amount of sunshine. Solar power can be used not only to produce electricity but also for other processes such as the 'sea-water greenhouse'. This is a highly innovative system developed to produce water to enable crop growth in arid regions. It combines three naturally occurring resources, sunlight, sea water and wind, to provide distilled water. The focus could be changed to water production, with crop production achieved as an additional benefit. An optimized plant could possibly produce water at the rate of 550 cu. metres per day per hectare.

Transporting water from distant locations has often been considered in the past. Tankering water from water-rich countries and towing icebergs from Antarctica have been suggested but not pursued for technical, economic or environmental reasons. Tankering is again being promoted, from North America, but environmental issues have still to be resolved. Icebergs should perhaps not be forgotten, as 60 per cent of the fresh water in the world is located in Antarctica.

One final resource is 'water mining'. This is the abstraction of ancient, or fossil, ground water, sometimes of poor quality, from aquifers where there is little or no recharge. Its use can hardly be termed 'sustainable' as it will clearly run out at some time. In terms of sustainability, the status of ancient water is similar to that of fossil fuels. Whether it can be termed a resource is questionable. At best, its use would delay the depletion of other aquifers which rely on annual recharge to maintain the system. If used, then clearly it has to be part of a strategy where other resources will be developed in time to replace the loss of the ancient water.

Water demand

Demand is the other half of the water shortage equation. Demand can be broken down into three main areas: domestic, industrial and agricultural. For the purposes of demand comparisons per capita we can assume that the domestic requirement is about 300 litres per day. Of this, only about 5 litres per day needs be of drinking-water quality. Industrial and service sector demand is unlikely to be above 75 litres per head per day. Agricultural requirements to meet the food needs of an individual are, however, much higher at around 3,000 litres per day.

Analysis of water demand in the Gulf indicates that agricultural water demands predominate, typically of the order of 70 per cent of water produced or abstracted, with domestic demand at 25 per cent and industrial around 5 per cent. These are only general figures, and the proportions obviously vary from country to country. The demand split does not take account of imported water in food, and hence does not reflect the suggested per capita requirements stated above.

It is clear from these demand figures, taken alongside the renewable resource of 250 litres per head per day, that the Gulf region cannot be sustainably self-sufficient. The rainfall available for aquifer recharge does not meet the required demand. On current understanding of the aquifers, many areas in the region are mining water to meet the water demand. There is already a water shortage; the debate is about when the resource will run out.

Water conservation

Conservation of water resources is a necessity, not an option. Current terminology defines this as demand management. In areas of the world where there is a water deficiency, demand management comprises two elements. The first is to ensure that the water is being allocated to the correct use and for an advantageous return. The second is to ensure that it is being used productively. Better to be doing the right thing a little badly than the wrong thing efficiently.

Water allocation

It is natural for any government to want to provide adequate supplies of water and food to the population. From the figures given for demand, agricultural use at 3,000 litres per head per day – ten times the domestic requirement – has to be examined in detail. With the growing populations in the Gulf area and limited water resources it is clear that self-sufficiency in agriculture cannot be achieved. The only way to balance the water resource/demand budget is to import water incorporated in food from areas of the world with a water surplus. This policy has enormous political implications, both in reducing current agricultural production and in accepting dependence on the global food market. It is increasingly believed that there is a shortage of fresh water to meet the world's growing demand for food. This means that Gulf countries must continue to play their part in ensuring that the world's water resources and environment are managed in a sustainable way.

Determining the rules for the allocation of water in the Gulf area is not easy. Which industries should take precedence? Should some crops continue to be produced, and if so which ones? To move away from the situation which is depleting the natural water resources to one which improves the water resource will require an economy that is diverse, flexible and robust. Obviously a reduction in agricultural activity could not be achieved overnight. Any remaining agriculture needs to be made as productive as possible.

Water productivity

Agriculture Traditional irrigation techniques lead to very significant losses in water resources, either upstream through evaporation in open channels or, in plots, through infiltration and loss into the ground of most of the water used. On average, worldwide, only one-third of the water supplied benefits plant growth. There are irrigation systems in the Gulf region that are state-of-the-art and about as economical as can be achieved. However, the overall efficiency of many irrigation systems, the ratio of water lost through evapotranspiration to water drawn, is generally not well known and controlled, and the potential for saving water is considerable. Water lost through infiltration does, however, recharge the downstream ground water, which may be of further use.

The management of irrigation systems requires decisions and interventions on a continuing basis. These cover the full system from distribution to application and down to the crop root zone. This in turn requires reliable and adequate information and data, the need for which was noted above. Evapotranspiration needs to be measured at local level to determine irrigation efficiency. However, because of water re-use, the efficiency at an area or basin level is far higher. Collecting the ever-increasing data imposes a burden on those operating the irrigation systems. New techniques such as Satellite Remote Sensing and Geographical Information Systems (GIS) are helping to obtain and handle data of this kind.

Domestic Domestic use is next in order of demand after agriculture. Many of the tools and techniques for reducing demand or re-using domestic waste water are not new, particularly to the Gulf region, which probably leads the world in the re-use of domestic waste water. Does the use to which the water is put need reviewing? More emphasis could be placed on agricultural use, perhaps with ultraviolet disinfection to further safeguard the public. Alternatively, there is the potential to use existing aquifers by using artificial recharge principles and effectively forcing recycled water into the aquifer under pressure to be re-abstracted elsewhere. This will allow the water quality to be improved within the aquifer itself. The use of artificial recharge does, however, present a number of problems, with possible chemical and bacteriological contamination of the native aquifer. Studies are required to determine the quality constraints of the recharge waters, recognizing the need for care in what is infected so as not to pollute the aquifer. The other aspect of artificial aquifer recharge is that the aquifer has to be of reasonable water quality. There is no point in losing water to a saline aquifer.

Artificial aquifer recharge happens in some urban areas by default. Water from septic tanks and water distribution systems causes waterlogging of the soil. In some areas it is used for garden watering. Water consumption will increase when sewerage and leakage prevention schemes reduce the waterlogging.

Preventing the leakage of water from the distribution system is always one of the first techniques to be mentioned. As with most techniques for saving water, its application is not as easy as it sounds. The first attempt to modernize leakage control in the UK was made in 1980 when *Report 26* was published by the National Water Council. The results were generally good while management was keen, but the distribution systems soon slipped back to their previous standard when resources were needed elsewhere. The methods described in *Report 26* were exported to the Gulf region, where again they were initially successful but frequently met a similar fate. One of the recommendations of the report was that a standard method be developed for the reporting of leakage and that the use of percentages of quantities supplied be abandoned.

In 1991 the UK set up another working group on leakage which reported in 1994. Much had changed with regard to the technology: data loggers, computers, telemetry and improved water meters were now readily available, automating and improving much of the data collection and analysis. The only thing that had not been resolved was how to report leakage. It is not possible to measure the performance of municipalities or water companies around the world as there is no agreed method. In Europe, recommendations are only now being made for a standard methodology to be used.

The principal components of the leakage calculation are the quantity into supply, the metered consumption and the unmetered consumption. Using the variation allowed by the water regulator in the UK on the accuracy of the measurements, it is possible for one water company to report losses of either 24 per cent or, if it played the numbers game to its advantage, 9 per cent.

The 1994 report considers the setting of an economic level of leakage, estimating unmeasured water delivered, interpreting and using night flow data, managing water pressure, dealing with customer leaks, and leakage management, technology and training. Pressure control is often highlighted separately from leakage control. In the Gulf region its potential for savings is generally small, given the flat topography, but its application should not be forgotten.

Dealing with customers' leaks has traditionally been the responsibility of the customer. Given the expertise within the water-providing bodies and the need to conserve water this may not be the best policy. Free repairs may save more than they cost to undertake. Where the customer is being metered there is obviously an incentive for the customer to locate and repair leaks, but this does depend on the cost of the water.

Metering all domestic customers is also advocated as a method of reducing demand. Its implementation and operation are not as easy as may at first appear. Most domestic water customers in the Gulf region are already metered and the issues of universal metering will be well known. Experience from around the world highlights the problems of reading and maintaining the meters, collecting the money and, most importantly, setting the tariffs. The comments here on metering are general and do not relate to any particular country. Some countries are running effective metering operations, but tariffs have still to be addressed.

The principle behind a metering policy is to have a system that provides for fair and affordable water charges, particularly for vulnerable customers, while ensuring the sustainable use of water supplies. It has to reflect the vital role which access to water plays in maintaining the good health of a nation. Even in comparison with other utilities, the service provided by water companies is essential to the well-being of customers and to the protection of public health. In the right circumstances metering can play a useful role in giving consumers messages about the value of the water they use, and the right incentives to economize in their use of it.

The effect of measured charges on different groups of customers depends on the way the tariff is structured. Traditionally, tariffs have consisted of a fixed standing charge plus a variable charge based on a single price for each cubic metre of water consumed. Tariffs of this sort are not particularly favourable for people on low incomes, both because of the standing charge and because water used for essential domestic purposes (such as drinking, cooking and hygiene) is charged at the same rate as water used for discretionary purposes (such as garden watering, using sprinklers or filling swimming pools).

An alternative structure of water tariffs which is fairer for those on low incomes would involve removing the standing charge and incorporating it in volumetric charges. Also favoured are more imaginative tariffs which provide stronger incentives for customers to economize on the use of water for discretionary purposes without discouraging essential use. These might include ‘rising block tariffs’, with a relatively low rate per cubic metre for a first tranche of water, and a higher rate for subsequent tranches of water. Another option would be seasonal tariffs designed to provide appropriate incentives for efficient garden use, with summer use above a certain level charged at a higher rate than winter use.

Certain households, particularly those with low incomes, will find water bills a particular burden. A range of payment options and other rights should be available to ease the difficulties faced by such households, including a choice of frequent payment options (e.g. weekly/fortnightly) without extra cost to the customer. Households on low incomes, particularly large families, and people with special medical needs could perhaps be given the option of a charge based on average household use, rather than their actual meter reading, if they wish to do so. This will protect vulnerable customers from high bills because of an unavoidably high use of water.

The main area in which metering for water differs from other services is that disconnection of household customers from the water supply is unacceptable. Water is essential for life and public health and should not be denied to households simply because of non-payment of bills.

A part replacement for domestic use is 'grey water'. This is waste water from baths, wash basins, washing machines, etc., which can be used for flushing toilets or watering gardens. Its implementation requires extensive modifications to existing buildings to collect the grey water, store it and then return it to the toilet-flushing cisterns. It does have potential in new buildings. Its other main drawback is that in areas where there is 100 per cent re-use of treated effluent the use of grey water reduces the quantity of effluent. If effluent is used for agriculture it may have to be replaced with clean water or the extent of the agriculture may need to be reduced.

Water used for flushing toilets makes up the largest element of water supplied to both domestic and office premises. Typically cisterns were of 9-litre capacity, but 6-litre cisterns have been available for some time. Dual-flush cisterns are also available. Their use needs to be encouraged. Taking a shower has always been considered to be better than taking a bath, using 30 litres instead of 80 litres. The increasing use of power showers, which use up to 120 litres in the same period, means, however, that care has to be exercised in encouraging the use of showers as a means of saving water.

Other water-saving devices available are spray taps, flow restrictors, the more modern water-efficient washing machines and dishwashers and, in commercial buildings, the use of automatic devices for the control of urinal flushing. All will help to save water, but will not have a major impact.

The adoption of drought-tolerant plants and drip irrigation of gardens reduces water consumption; these practices are well understood and encouraged in the Gulf region.

Savings of high-grade water (drinking water) have been made in some countries by providing low-grade water or sea water to consumers for use in flushing toilets. In Hong Kong, for example, sea water has been supplied since the late 1950s, primarily for flushing, in government and government-aided high-density development schemes, and subsequently for flushing supplies throughout the urban areas and the new towns. Over 70 per cent of the population are now supplied with sea water for flushing. Sea water is not treated to the same standard as fresh water but its standard still has to comply with the guidelines laid down. The sea water is first screened and then disinfected with chlorine or hypochlorite before being pumped to service reservoirs for distribution to consumers. In the Gulf region the benefits of adopting this would have to be examined very carefully, as the effluent produced would change in quality and be less useful. There is also the not inconsiderable cost of the second distribution system. The use of sea water in large cooling/air-conditioning systems is another application to be considered.

Industrial The final demand group is the industrial user. It goes without saying that high water use industries are not particularly suited to this region. However, industry should be checking on its water use and carrying out water balances. Producers of high-value products do not always appreciate or consider important the amount or cost of the water they use. Typically, the highest consumption is in cooling and steam production. Measures for improving the efficiency of water use can be divided into three areas: water management techniques, good housekeeping, and plant and process modifications.

Water management techniques basically involve applying the old adage 'if you don't measure it, you can't manage it'. Measure the water used in the process, understand how much is used and where, compare actual consumption with theoretical consumption. Then identify the changes to be made to optimize water use.

Good housekeeping means implementing leak detection programmes (if the supply authority will not), putting triggers on hoses, reducing cleaning needs, and undertaking valve and pipe maintenance. Plant and process modifications are best introduced when investing in new plant, when efficient water management and environmental protection systems can be incorporated into the original design. These include items such as recirculation of cooling water, improved cooling towers/systems, in-process water recycling, re-use by cascading water from cleanest to dirtier uses, and improving plant and vessel washing techniques.

Implementation of water conservation and sustainable water use

Very few of the ideas discussed, many of which are not new, are likely to be successfully implemented if the correct institutional and economic frameworks are not in place. The adoption of demand management policies requires political maturity, with difficult decisions needed now for long-term gains to be enjoyed by the next generation. Not all actions will be received easily by the public, but the communities' support of the actions and policies needed to achieve a water balance is vital.

The main potential barriers to the wide-scale adoption of these water conservation technologies and measures include the following:

technical barriers:

- the need to adapt technologies and techniques to specific applications;
- insufficient information on field performance;
- non-conventional physical characteristics;
- uncertain effects of waste transport and treatment;

social barriers:

- actual or perceived adverse reactions from users;
- a lack of awareness by key end-users of the potential for water conservation;
- general consumer resistance to change;

economic barriers:

- the under-pricing of water in some regions;
- the higher installation cost of water-management devices;
- limited government subsidies for the introduction of water conservation technologies/measures;

- large investments in process technologies with high water consumption which have already been made (and have not yet been depreciated);
- the absolute scale of the re-investment required to switch to water-efficient processes;

regulatory barriers:

- the lack of suitable standards and regulations (notably water quality requirements in irrigation, and building regulations encouraging the installation of water-conserving devices).

Important activities to overcome these barriers to water conservation measures include:

- the instigation of various innovation tools (for end-users and water suppliers alike) such as demonstrators, education/awareness (including use of mass media, personal contact and special events) and decision support tools;
- the introduction of appropriate economic and financial instruments, such as taxes, subsidies and charges/tariffs;
- the development of appropriate regulations and standards such as restrictions, allocation/licensing and building/planning regulations.

Institutional issues to be resolved to achieve global and sustainable water resource management range from global and regional planning of urban and rural development to community management of water services and irrigation. Actions to consider include:

- acceptance of the view that, as two-thirds of the world's major catchments are shared by several countries we must move away from each country managing *its* water on *its* territory separately;
- the introduction of integrated information systems to bring together data held by different sectors – agriculture, industry, public works, development, tourism, environmental protection, etc.;
- an improvement in the regulation of the water allocations between the various users;
- establishment and enforcement of regulations for withdrawals and discharges of water;
- provision of access to information and raising the awareness of

responsible local officials, user representatives and the population in general;

- creation of conditions for financial viability of projects and schemes: water management must integrate social and environmental elements, which makes investment and strategies more complex.

The establishment of one agency whose sole responsibility is the protection of the water resources is probably the only way to achieve progress towards a sustainable water policy. The agency must be independent of the agricultural, municipal and other water users and have the full support of government. Tariffs for water abstraction need to move towards reflecting the true cost of the impact of depleting the water resources. Water users need to be made to think twice about the amount they use and the uses to which water is put. Inappropriate industries will be discouraged, and there will be added incentives to find alternative water sources. An important role for the agency would be the auditing of the performance of water supply companies and irrigation schemes in areas such as water quality and leakage.

Privatization of the water supply operation, as currently being considered in the UAE, if set up with the correct incentives, can be an effective way of improving efficiency and delivering water use reductions.

Public awareness and appreciation of the water resource issues locally and globally will have the greatest impact on achieving sustainability. Education and involvement of the population in local decisions is going to be needed. The public must want to change.

Conclusion

There is a growing awareness that the world's water resources are coming under threat. The concerns for the Gulf are possibly more immediate. The need for accurate data on all elements of the water cycle is being constantly emphasized by all parties attempting to monitor the situation and devise strategies to cope in the future. The same data also have uses at country level to enable improved management of resources. Active involvement through international agencies on issues concerning global food production is important for countries in arid regions, which need to import water as food. The successful introduction of water conservation techniques and the education of all users and providers of water is essential to help conserve water. Concerns over CO_2 production will impact on the use of desalinated water which goes some way to making up for the current deficit in the water resources. Therefore the development of alternative sustainable resources becomes a necessity.

7 Managing the oil spill risk

Brent Pyburn

The discharge of oil into the waters of the Gulf states can have a traumatic effect on their fragile ecosystems. Although a major release of oil into the water has an immediate effect, the consequences can be limited to the short term if a rapid and considered response is carried out.

While the major concerns have always concentrated on these larger releases of oil, the frequent small releases as a result of ship or terminal operations can have a more chronic effect, particularly on the local environment. This is particularly relevant in the Gulf states where there is a large concentration of terminals and tanker traffic.

The key to dealing effectively with the risk of oil spills is to have measures in place to reduce the risk of their happening and, if they should happen, to respond to them effectively. This paper examines the benefits of ensuring that the standards of ships and their crews, and the standards of operations at oil terminals, play a significant role in reducing the risk of oil spills happening. It also recognizes that the human element is an unpredictable factor in the cause of accidents and that although preventive measures are put in place it is prudent to have a response mechanism in place. These response arrangements must be effective and sustainable. The paper emphasizes the need for good planning, identification of appropriate resources and cooperation with all the parties that would be affected by the spill.

Finally, the paper looks at the changing philosophy regarding response strategies and in particular the need for a more considered and sensitive approach to clean-up.

Prevention

Oil may be discharged into the water, either deliberately or accidentally. In the majority of deliberate releases the motive is simply commercial. The vessel may have oil-contaminated ballast or tank washing water and has to get rid of it to enable it to carry more cargo. Alternatively, the owner of the vessel has reduced the operating costs of the vessel to such a level that the vessel is badly managed and equipment critical to the safe operation of the vessel is faulty.

International conventions have been established which make it a criminal offence for a vessel deliberately to discharge oil or oil-contaminated water and other noxious substances into the sea.

International conventions

The main convention dealing with oil pollution is the International Convention on the Prevention of Pollution by Ships, more commonly known as MARPOL. The convention outlines practices and procedures for the construction and operation of vessels, with a primary focus on preventing the discharge of oil or water contaminated with oil above a set level of parts per million. It also outlines the responsibilities of ports to provide facilities for the reception of contaminated water that constitute a further incentive for shipowners and shipmasters to act responsibly. Other conventions include the International Convention on the Prevention of Marine Pollution by Dumping (the London Dumping Convention), which focuses on the prohibition of dumping of hazardous materials and 'permit to dump' requirements; and the International Convention Relating to Intervention on the High Seas in Case of Oil Pollution Incidents (the Intervention Convention) which is focused on governments' powers and responsibilities for intervening in the event of a casualty.

Policing operations

Unfortunately, however, there are a number of shipowners and shipmasters who will flout the rules of the conventions for commercial reasons. An effective way of dealing with these illegal releases is to police the traffic lanes using an aircraft fitted with monitoring equipment. Unscrupulous shipmasters are extremely nervous about deliberately discharging oily water into an area that they know is being monitored. The UK government carries out regular patrols over the busy shipping lanes of the English Channel using a twin-engined aircraft. The monitoring equipment fitted to the aircraft consists of infrared, ultraviolet and side-looking radar that allows monitoring to be conducted round the clock: while the aircraft is not, of course, flying twenty-four hours a day it may stagger its patrol times so that vessels transiting the channel are never sure when it may be overflying them. This can be a very successful deterrent.

Preventing accidental release

Where accidental releases occur, in the majority of cases the reasons are human error, structural failure of the vessel or mechanical failure of critical equipment. In all of these cases the risk can be greatly reduced by operating a marine assurance policy of vessel selection and auditing of terminal and operations.

The BP Group policy

The BP Group Health, Safety and Environment Policy established in 1996 reads: 'Our goals are simply stated – no accidents, no harm to people, and no damage to the environment.' In addition to the Group statement, the company's annual report and accounts states: 'We aim to carry as much of our oil as possible in our own ships.' By carrying oil in its own vessels BP is directly in control of managing the risk of an accident happening. The international fleet of 26 owned and managed tankers is maintained to a very high standard and all are manned by BP-employed officers. The fleet has recently been boosted with the bareboat charter of two modern double-hull very large crude carriers (VLCCs) and three new 100,000-tonne vessels.

Although much of the oil supplied into the BP system is carried in the company's own vessels, the need to optimize the use of our vessels and the flexibility demanded for traded oil means that we are still a major user of voyage chartered vessels. These are obtained through the 'spot market' and in 1998 amounted to approximately 1,500 voyage charter movements.

Any marine venture carries an inherent risk with it. That must be managed by carefully and scrupulously assessing the risk and then having the necessary processes in place to manage it. The BP Group carries out this risk assessment and management process through its shipping policies, which are implemented and audited by the BP Shipping Risk Management Group. BP Shipping also offers its expertise in risk management to third parties.

BP Group shipping audit policy

The company also recognizes the risk at the beginning and end of each voyage, and puts a substantial effort into managing the risk at the marine interface of oil terminal operations. This was originally focused on the operation and physical features of the jetty or berth; the scope has now been increased to cover the operation from picking up the pilot inwards, to dropping off the pilot outwards. It therefore now includes, for example, river and estuary transits, where a vessel carrying BP's oil destined to or from its facilities could face the greatest risk. Included in the terminal audit is a review of the emergency response procedures and a check on the resources available to respond to an oil spill.

BP Group time charter policy

There is also an increased commercial and reputational risk to the Group in engaging time charter tonnage over that of spot fixtures. Among ways of managing this risk is the requirement for the organization responsible for the daily management of the vessel to be audited.

BP Group ship vetting policy

There is also a policy on managing the risk of using spot or voyage charters. This policy is based on a long and continued operation of BP's own tanker tonnage as well as many thousands of chartered vessels. We have specified certain operational practices and policies over and above international statutory requirements as part of our approach to risk management. These are as follows:

- Toxic or volatile cargoes must be closed loaded, regardless of the allowances made in some industry codes.
- Tankers that are fitted with an inert gas system must use it when carrying any BP cargo.
- On chemical vessels, the master must have a minimum of two years' chemical shipping experience, and the chief officer and chief engineer must each have a minimum of one year's chemical shipping experience.
- All vessels involved in a cargo ship-to-ship transfer operation must be assessed as 'acceptable for BP Group use' through the vetting system.
- Combination carrier type vessels in excess of 15 years of age must not be used.
- Owner or manager assessments will be carried out where we are exposed to increased risk through high vessel use (in, for example, time charters), or if the fleet profile is poor in respect of, for instance, poor inspection results, terminal feedback or casualty reporting.

Careful ship vetting provides a procedure for ensuring that, as far as is practicably possible, a vessel used for transporting oil will comply to a particular standard. It should be made clear here that the vetting of a vessel does not just mean that the vessel is inspected. Vessel inspection is a significant part of the vetting process, but there are other considerations that need to be taken into account when assessing the vessel.

Shipowner's reputation

Careful consideration is given to the reputation of the company owning the vessel. This is assessed by using intelligence from market sources. The reputation of the company can be a crucial factor when deciding whether to accept promises of corrective action should a fault be discovered on one of its vessels.

Owner audits

This audit establishes whether the owner has the necessary management processes in place to manage ships effectively and safely, with consideration for the environment, and whether it has the organization and resources in place to respond speedily and effectively to an emergency.

Ship inspections

Individual ships are inspected to ensure they meet a certain standard.

SIRE and CDI reports

There are two industry databases where inspection reports on vessels are filed: the Ship Inspection Report or SIRE, operated by the Oil Companies' International Marine Forum (OCIMF), and the Chemical Distribution Institute database, known as CDI. Individual ship reports can be extracted from the database, which may forestall the requirement to inspect a vessel individually.

Feedback

All BP terminals are required to submit feedback using a questionnaire on all vessels using their facilities. Similar feedback is required from the charterers on all vessels fixed on a BP account.

Flag changes

Changes of flag, particularly if associated with a change of class, would cause a general review to be carried out on the vessel and owners. The review would look at the reasons for the change, the vessel's age, its special survey and docking cycle.

Market intelligence

Using a number of sources such as publications and industry forums, market intelligence can identify problems with shipowners or individual vessels. Port State Control Detention Lists are scrutinized.

Casualty data

The company pays an annual fee to Lloyd's of London, which reports casualties or other serious events to their customers. If such a report is received on a vessel used by the company it is referred back to the owner of the vessel in question.

Manning and management changes

The company is very conscious of the role of the 'people factor' in contributing to accidents. For this reason any change to manning on the ships or management in the owners would cause the vessel's status to be reviewed.

Change of class

The reason for any change of class needs to be understood. If it is a move to a non-member of the International Chamber of Commerce or if it coincides with a significant event such as damage repairs or special survey, it would cause the vessel's status to be reviewed.

By carrying out these comprehensive risk management processes BP can minimize the risk of an accident happening. BP actively encourages the oil and shipping industries to follow similar risk management policies. Experience has shown that the whole industry suffers when an incident occurs, particularly a major oil spill.

However, no matter what risk mitigation procedures are established and maintained, the 'human element' in any accident is difficult to manage simply because of its unpredictability. In the recent most notable incidents, such as those involving the *Exxon Valdez* and *Sea Empress*, the accidents were caused by highly trained professional mariners, on relatively new and sophisticated vessels, not following strictly laid down procedures that were established precisely to prevent such incidents from happening. It is for this reason that, no matter how good the policies designed to prevent accidents, there is still a need to have processes in place in the event of an accident happening.

Oil spill response

The most effective tool for responding to an oil spill is the contingency plan. No matter how much equipment you have and no matter how expert your people are, unless there is a plan to deploy those resources effectively then the response will fail.

The scope of the plan should deal not only with the 'sharp end' tactical response to the incident but also the 'softer', more strategic issues such as reputation and 'licence to operate', and how these issues are managed.

Resources are also a critical element of an effective response. While emphasis is always put on specialist equipment and expertise, the response effort requires a wide range of skills in such areas as logistics, staff management, diplomacy and communications.

In BP our attitude to oil spill response is one of over-reaction, assessment and de-escalation. If we are committed to responding, then we will commit all the necessary resources required to complete the response in as short a time as possible, with as little further damage to the environment as possible. This is particularly relevant when considering response strategy, and is discussed below.

Contingency planning

As mentioned earlier, the oil spill contingency plan must cover both the tactical and strategic elements of a response. It also has to cover a wide range of scenarios from the small spill to the catastrophic spill – the latter being the kind of event we associate with something like the *Exxon Valdez* incident.

The plan itself is a living document that needs nourishment at regular intervals. It is pointless producing a document that is never reviewed or updated and just sits on a shelf, or is used as a prop for keeping a door open! (In my travels around the world I have seen this on a number of occasions.) There is nothing worse than referring to a plan where the data in the plan are out of date.

A successful contingency plan has to be produced in a structured way and contain all the elements required to carry out an effective response. These elements include the following:

- *Ownership of the plan.* It is essential that someone is identified as the owner of the plan, who has the responsibility of keeping the plan as a living document.
- *Scope of the plan.* The scope of the document must be clear. For example, what are the geographical parameters of the plan? What other plans integrate into this plan? What local, national and even international regulations or requirements drive the plan contents?
- *Risk assessment.* An assessment of the risks appropriate to the parameters of the plan must be carried out, taking into account, for instance, the type and amount of oil liable to be spilt; the navigational hazards associated with entering and leaving the terminal; the fate and effect of spilt oil, taking into account the weather, tidal peculiarities and environmental sensitivities of the locale.
- *Scenario development.* On the basis of the known risks, a wide range of scenarios should be developed covering all the operations carried out within the parameters of the plan.
- *Response strategies.* Strategies should be developed for managing the various scenarios.
- *Response management.* An organization must be put in place to manage the response strategy.

- *Resources.* Resources must be identified to service the response management and provide hardware for the physical response.

The importance of the national plan

During the late 1980s and early 1990s a spate of oil spill incidents, including the *Exxon Valdez* and *Haven* incidents, showed that the relationships between the oil industry and federal and state governments, and between governments of adjoining countries which shared a coastline, were crucial in expediting an effective response. In the case of the *Exxon Valdez* the relationship between the oil industry, the state authorities and the Prince William Sound communities was abysmal. In the case of the *Haven* incident, where oil was spilt in Italian waters and polluted beaches in both France and Italy, the confused response starkly showed the lack of coordinated planning between the two countries. The national plan is crucial in coordinating an effective response to a major incident. Only governments can organize the logistics and legislation that may be required to bring in equipment and expertise speedily from international sources. As we have seen, oil spills do not respect international boundaries and it is important to consider your neighbours when planning for an effective response. Cooperation between the government and the oil industry in the country is also an important factor in ensuring that the best expertise and equipment are available. It is likely that the oil industry in the country is the best source of expertise and ready funds to provide assistance to a government response.

The International Convention on Oil Pollution Preparedness and Cooperation

Both the oil industry and governments, through the auspices of the International Maritime Organization (IMO), recognized the importance of the national plan in underpinning a coordinated response. During the 1990s the industry, through the International Petroleum Industry's Environment Conservation Association (IPIECA), has worked together with the IMO to improve response capability worldwide. In 1990 the IMO adopted a convention called the International Convention on Oil Pollution Preparedness Response and Cooperation 1990, more commonly known as OPRC 90. The focus of the convention was on national preparedness and multilateral cooperation, and the main provisions which signatory countries were required to observe were:

- the designation of a competent national authority with responsibility for oil pollution preparedness and response;
- the designation of a contact point where notification of incidents could be received and appropriate response action taken;
- the identification of an authority entitled to act on behalf of the state in providing a response, such authority to hold the necessary resources and have the organizational structure to manage a response;
- a national plan which would be the foundation of the national response;
- international agreements with neighbouring countries and other states to cooperate in the unhindered movement of resources across international boundaries to the scene of the incident;
- bilateral and multilateral cooperation with the oil and shipping industries and other interested bodies in providing a coordinated response to an incident and providing technical advice in raising the capacity of a country to respond.

OPRC 90 became the focus for a joint oil industry and IMO initiative to raise awareness of the need for contingency planning. In late 1991 a joint IPIECA/IMO seminar in Jakarta set the scene for a series of regional seminars on the theme of cooperative contingency planning. One of these seminars was held in Bahrain in 1993 for the Gulf region. The seminars brought together senior government and industry representatives to explore the contingency planning process and develop cooperation in planning. They helped to bring government and industry closer together and to focus government thinking away from hardware towards a realization that good management and organization were the basis of effective response. During the seminars the profile of OPRC 90, the Civil Liberties Convention (CLC) and Fund Convention (international compensation regimes) was raised and some countries were persuaded to sign the conventions. The 'tiered response' concept of planning (discussed in detail below) was also promoted as the preferred structure on which to build a plan. This was a big step forward.

Joint IMO/IPIECA Global Initiative

As a follow-up to the series of seminars, the IMO with IPIECA developed the Global Initiative (GI). This was a programme designed to provide assistance to developing countries in planning for oil spills. The initiative has produced many positive actions, including the following:

- on invitation, a technical mission has visited the country concerned and provided a forum for government and industry to debate the development of the national plan, in some cases establishing the process through which the plan is developed;
- holding workshops for training government officers in environmentally sensitive area mapping – a crucial element of any national plan;
- establishing sponsorship for projects from the World Bank, Global Environment Facility and the UK and French governments;
- developing with the oil industry IMO modular training courses on oil spill management;
- carrying out joint IMO/industry training courses in Africa and the Asia-Pacific region.

The 'tiered response' concept in planning

Increasingly governments and industry are using the 'tiered response' concept when planning for oil spills. The concept provides a cost-effective method of ensuring that there are appropriate resources and an appropriate organization in place to respond to a wide range of scenarios. By carrying out a risk assessment for each tier, the correct type of equipment and the optimum amount can be determined.

A *Tier One* plan is designed to cover the small operational spills at a facility or terminal. These spills can be caused by a transfer hose break, a tank overflow or other similar occurrence. The organization in place to respond to this type of incident must be able to respond within an hour, with sufficient appropriate equipment to deal with the type and likely amount of oil. To provide such a speedy response the organization and equipment must be located close to the area where the small operational spills are likely to occur.

A *Tier Two* plan is designed to cover those incidents where the resources at the terminal are insufficient to respond effectively. It may be that the amount of oil is greater than the response capacity of the equipment or that the location of the spill is away from the Tier One resource. In this case other resources need to be brought in. In some cases Tier Two spills are serviced by a mutual aid organization combining the resources of a number of companies in the area, or through a stockpile of equipment funded by a number of interested companies.

A *Tier Three* plan is designed to cover worst-case scenarios where not only national resources are used, but assistance is sought from international sources. The Gulf Area Oil Companies' Mutual Aid Organization is a good example: this provides, in the event of a catastrophic incident occurring in the Gulf area, for oil companies in the region to donate equipment and expertise. Certain oil companies also contribute to international response centres, which because of their prime response function are called the International Tier Three Centres.

International Resources: The International Tier Three Centres There are a number of Tier Three Centres around the world. The international oil industry funds centres in Florida (Clean Caribbean Cooperative), Singapore (East Asia Response Pte Ltd) and Southampton, UK (Oil Spill Response Ltd). The Petroleum Association of Japan also funds a number of small centres in Japan and on the major sea route from the Gulf to Japan. The three main centres funded by the oil industry all work to similar principles of providing state-of-the-art equipment accompanied by appropriate expertise. The service is provided on a non-profit-making basis. Each of the centres has systems in place to mobilize resources rapidly and airlift them to the scene of the spill.

One of the main advantages of these centres is the emphasis they place on equipment maintenance. There would be nothing worse than flying equipment to a spill thousands of miles away and finding it does not work when it arrives, or fails during its first day of deployment. Therefore the centres each have a comprehensive planned maintenance system.

Another advantage is the core of dedicated personnel who staff these centres. They are highly trained and know their equipment intimately. However, they should not be seen as just equipment operators. They have skills in supervising response teams and advising response managers on both tactical and strategic responses. During the *Exxon Valdez* incident a small team from Oil Spill Response Ltd played a significant role in supervising shore clean-up crews, training new responders and advising the response management on strategy. They have since done this on a number of major incidents.

It is important that the role and responsibilities of these centres in an oil spill are clearly understood. The centres are there essentially to support a response operation. They will not manage the overall response, but may provide levels of supervision for certain aspects of the response operation. They are a valuable source of specialist equipment and expertise.

Response strategies

Over the last few years, as a direct result of experience gained in Alaska, and the *Braer* and *Sea Empress* spills in the UK, there has been a radical change in response philosophy.

During the *Exxon Valdez* incident it was calculated that less than 3 per cent of the oil spilt was actually recovered during offshore containment and recovery operations. Some of the oil evaporated; the rest came ashore. As a consequence of a very aggressive shoreline clean-up strategy of hot water pressure flushing, severe environmental problems were caused by killing off the species of flora and fauna inhabiting the shoreline. In some cases such measures as were taken were necessary to protect the seal pups, birds and sea otters that were using certain stretches of shoreline; in other cases, however, such aggressive cleaning was not necessary and was a cosmetic exercise.

During the same incident the use of chemical dispersants was prohibited even though in the initial stages of the incident the oil was still close to the stranded vessel and in relatively deep waters. The correct use of dispersants at that time could have had a profound effect on the whole response operation with little or no impact on the marine environment.

The *Braer* incident illustrated to the world that nature itself can provide the most sensitive and effective response to a major oil spill. In 1993 the *Braer* foundered in very heavy weather off the west coast of the Shetland Islands. Fortunately the crude it was carrying was very light in nature and the very rough weather that had put the tanker ashore enhanced the break-up of the oil into very small droplets. These droplets dispersed into the water column and allowed speedy biodegradation of the hydrocarbons.

In the *Sea Empress* incident, the British government, following its primary response strategy of using dispersants, effectively reduced the amount of oil coming ashore by early and selective application of the chemical. For the first time in a major spill, a team of specialists was contracted in to monitor the response operation and in particular the effectiveness of the dispersant spraying operation. The results showed clearly that the use of dispersant played a significant role in the effectiveness of the overall response.

The problems caused by inappropriate mechanical response strategies and the recognition of the successful dispersant operations carried out in recent years have caused a rethink in the world of oil spill response. Dr Jenny Baker, a leading environmental consultant, has persuaded the oil industry to adopt a more considered approach to the issue by using net environmental benefit analysis (NEBA) before deciding on a particular strategy. NEBA is designed to enable selection of a response strategy that will have minimum further impact on the environment. For instance, clean-up on a rocky shore on a coastline exposed to severe weather may be left to nature itself. Similarly, clean-up of a salt marsh by physical removal of the oiled growth may have a more traumatic effect on the marsh than letting the oil biodegrade naturally. Dr Baker also encourages the use of dispersant, where the net effect of using the chemical is of clear benefit to the environment. For instance, the use of chemical dispersants over corals may be more beneficial than allowing the oil to be stranded on the roots of a mangrove. A key element in the use of NEBA is having someone on the spot who has the necessary skills to make a judgment.

The oil industry recognizes that there are countries in the world which either completely prohibit the use of dispersants or control their use to such an extent that their application is of little use. For dispersant to work effectively the oil has to be as fresh as possible. When oil hits the water a process of evaporation and emulsification called 'weathering' takes place. This process increases the viscosity of the spilt oil, making it less amenable to the use of dispersant. Consequently the window of opportunity for using dispersant is limited, and bureaucratic procrastination can further narrow the band of time in which such action is of most use.

At present the industry is making an active effort to persuade governments of the benefits of the correct use of dispersant. Nevertheless, a response strategy needs to be balanced and to take into account situations where mechanical containment and recovery may be a more appropriate option.

8 Commentary and discussions

Peter Kassler

Sustainable development

The overview outlines some reasons why sustainable development is so important to the United Arab Emirates and some of the choices and trade-offs required between the satisfaction of present needs and provision for future generations. The choices will be particularly difficult if climate change policies, in effect, ask the UAE and its neighbours to forgo some of their current development in the interests of future generations everywhere, including those in countries which have had much longer to develop than the UAE.

This concluding chapter comments on the main issues raised in the papers in the volume as well as in the very rich discussions at the seminar (these discussions are reported in italics).

Roger Rainbow's paper underlines the importance of economic development as the principal engine which can help to drive societies towards benign solutions of such dilemmas. There are large qualitative and quantitative differences between the sustainability needs of, say, a sub-Saharan African who might be happy with clean water, sufficient fuel and a reasonable life expectancy and those of a citizen of Amsterdam, concerned about climate change and rising sea level.

Although the Gulf states are very wealthy in terms of GDP per capita their economic development is not yet similar to that of Western economies enjoying similar per capita GDP. The latter will have received from earlier generations a rich heritage of industrial, social and intellectual capital (including industries, civil institutions, and scientific and technical expertise). As long as they are well governed (which is of course not always the case), this heritage provides the industrial economies with the resources to resist internal and external shocks robustly and to adapt flexibly to major change.

There are, however, advantages in belonging to a young society like the UAE, a major one being the ability to develop new models of sustainability adapted to local needs and avoiding some of the mistakes of those who have trodden this path before. Assisting this development is the huge amount of information now readily available on the Internet and elsewhere about virtually everything and everywhere under the sun. This places much more 'knowledge power' in the hands of relatively small nations, making their rulers and scholars aware of the development options available to them and of the successes and failures experienced in other countries.

One of the most important choices of this kind relates to a nation's attitude to 'globalization'. Today ideas, money, and technical and managerial expertise have a global reach. Satellite television and computers show Gulf residents the same images of American-style consumer society as are seen in England or France. Italian or Canadian managers can offer their competing services for a position in Dubai. To a large degree (but not with complete freedom, because of the constraints imposed by national sovereignty), Western corporations can decide whether to invest in Europe, China or elsewhere. These examples illustrate what might be called the 'opportunity' side of globalization, the availability of choices and options originating in other parts of the world.

There is, however, also a darker side. Climate change is a global development, and there is nothing that one country can do to isolate itself from it. So, also, is a great deal of pollution. Atmospheric pollution follows the wind and waterborne contamination moves around with currents and tides, again often from one country to another. International political and economic pressures can be transmitted from one country to others (by means of boycotts and embargoes, for example), often for capricious reasons and with unintended consequences. There are also issues of economic and political power determining the extent to which countries wish to open their economies to multinational investment with which domestic firms may not be able to compete.

Here, Roger Rainbow's advocacy of the 'triple bottom line', whereby a company will be judged not only by its profit performance but also by its contribution to social and environmental well-being, is a crucial new element.

In discussion the question was raised (with some scepticism) whether, for a multinational, the interests of shareholders and managers are aligned with those of society at large. Will CEOs' remuneration be based, at least in part, on their companies' social and environmental performance? If, on the other hand, major companies can assert and demonstrate their commitment to the triple bottom line, this ought to reduce many of the concerns of campaigners for sustainable development about the motivation and behaviour of companies like Shell.

Rainbow's conclusion is that globalization and global companies favour sustainable development, both in spreading awareness of environmental and resource questions requiring attention and in making available the world's best expertise and technologies to deal with them. He recognizes, however, that in the long run economic and political sovereignty cannot remain untouched in the face of globalization. This will remain an issue in the Gulf, as it is in the United Kingdom and elsewhere.

Global climate change

The international context

Michael Grubb's paper charts the prolonged slow progress of the global warming issue from scientific curiosity in the early nineteenth century to worldwide concern today. The way in which international consciousness about climate change was raised, initially by a small group of zealots, and then adopted all round the planet by a critical mass of environmentally concerned civil servants, politicians and diplomats, is unique in the development of a single issue.

The critical question today, and one which may deeply concern Abu Dhabi and other oil exporters is: what happens next? As a single issue, climate change has to take its place among other more traditional issues and policy areas which have hitherto competed for politicians' attention: economic growth, trade, defence, energy security and voters' love affair with the motor-car. In some countries, such as the low-lying small Pacific island states which feel their very existence potentially threatened within the next few generations by rising sea level, there is no contest. This is the number one issue. However, these are not powerful countries, nor are they among the principal emitters of greenhouse gases. Countries belonging to both the latter categories, principally the United States, rank climate change much lower among their concerns and are less inclined today to take robust action to reduce their domestic energy consumption.

Michael Grubb leads us even-handedly through these ambiguities. He describes the growing quality and confidence of the scientific research documented in the IPCC reports of 1990 and 1995 and the slowly grinding diplomatic and institutional process leading up to the signature of the Kyoto Protocol in late 1997. Kyoto produced international commitment by the industrialized countries to reduce their greenhouse gases by 5 per cent below 1990 levels over a commitment period from 2008 to 2012. The commitment is a differentiated one, the United States opting for a 7 per cent reduction and the European Union 8 per cent, while Australia commits itself not to increase emissions by more than 8 per cent.

The Kyoto negotiations, however, also reflected the fact that, five years after the Rio summit, CO_2 emissions generated by a number of major industrial nations were still on an increasing trend.

In the Kyoto process, the United States led initiatives to introduce 'flexibility measures' into the Protocol. These are market-based instruments which include provisions to trade part of a signatory's emissions with other industrialized nations. These include the so-called 'economies in transition' – Russia and the other former Soviet republics whose economies (and CO_2 emissions) collapsed after the break-up of the USSR and which therefore became potential major emissions-trading partners with the United States and other heavy consumers of energy. Other flexibility measures allow countries to obtain emissions credits for emissions-reducing projects executed by one country in another member of the same group and, under carefully controlled conditions, in partnership with developing countries. The economic argument for these activities, emphasized by Grubb, is that emissions can be reduced at the sites where this can be done at the lowest cost per tonne of CO_2. In practice, the trading arrangements will allow CO_2 reductions to be made (or appear to be made – there are still major problems of compliance and verification to be resolved) by closing down obsolete coal- or lignite-fired power stations in eastern Europe, rather than by energy economies in the United States.

Clearly, however, the complexity of these arrangements does not make for transparency, and the process of reducing greenhouse gases has diverged widely from the simple reduction of domestic emissions by quantifiable energy management measures. At the extreme, in fact, industrial countries can go a long way towards achieving their targets without taking any substantial action within their own frontiers. The jury is still out on the question whether a total absence of domestic CO_2 abatement achievements will be acceptable to the international community. However, the ultimate sanction of refusing to ratify the Protocol is still open to participant nations until they are satisfied with the arrangements agreed.

All this makes it extremely difficult to judge what Kyoto, and the process which it has set in motion, mean for Abu Dhabi and other oil exporters in terms of real reductions of oil sales compared to some baseline representing 'business as usual'.

At a formal level, the Kyoto signatories are committed to minimize adverse social, environmental and economic impacts of compliance with the Protocol on vulnerable groups of developing countries. This provision, and its predecessor in the Rio Convention, were very important to OPEC member states (which consider themselves to be vulnerable developing countries), and its retention ultimately gave them enough reassurance to allow them to sign the Protocol. The evaluation of these impacts is taken up in Peter Kassler's paper, commented on below.

Grubb closes his paper with a fairly cool appraisal of the results to date of the greenhouse gas management process. Some progress in abatement has been achieved by the European Union (for reasons which have little to do with climate change) and Japan, but the United States, Canada, New Zealand and Australia are far off target. On the positive side, however, a long-term energy management process has been set in motion, together with the institutions to oversee it. He concludes that action at present may be quite modest but will probably be strengthened over time, and that the most likely outcome will be a wide range of market-based instruments to control emissions.

Impacts on energy exporters

From the first appearance of global climate change in international political debate in the 1980s the energy-exporting countries have been concerned about the impact of CO_2 limitation policies in energy-importing countries on the demand for what in many cases is their only export product. They feared that this would reduce the commodity prices of oil, gas and petroleum products, with adverse revenue and GDP consequences for exporters. Also, since for a number of years carbon taxes were seen as the main regulatory mechanism, there was a fear that the burden of these would be added to the cost of energy exporters' imports from industrial countries. As 'green' legislation forced renewable forms of energy into consuming markets, oil-exporting countries were further concerned that reserves of oil and gas (again, generally the only resource possessed by energy exporters) would be devalued, and might ultimately remain unused in the ground.

Peter Kassler's paper investigates the validity of these concerns. The energy exporters (defined as countries with net energy exports exceeding $2 billion in 1996) are a fairly diverse group, extending from Australia at one extreme to countries such as Angola and Congo at the other. The core of this group, however, consists of seven Gulf states (Saudi Arabia, the UAE, Iran, Iraq, Kuwait, Qatar and Oman), all principally oil exporters and all having economies which are very dependent on oil revenues.

The vulnerability of these economies to commodity prices was on display at the time of our Abu Dhabi seminar in October 1998, when the price of Brent crude was around $10/bbl. This was the result of the producers' miscalculating Asian demand growth and had nothing to do with environmental measures. However, the effect of the consequent revenue shortfall and the concerns it raised provided a graphic illustration of what a 'greening' world might really be like for major oil exporters. The Brent price has since recovered, but the problem for the exporters has not gone away.

As suggested above, the Gulf economies are not fully defined by their per capita GDPs, which tend to be high when commodity prices are high and much lower when they are not. They have particular development requirements, largely related to their undiversified economic structures:

the non-oil elements need to be strengthened to increase their resilience to shocks (including both normal business-cycle commodity downturns and environment-related events). The Gulf countries also have young populations needing education and employment, as well as security concerns leading them to spend considerably more than the global average on defence goods.

At an early stage in the climate change negotiations, some OPEC countries saw the process as an opportunity to obtain monetary compensation from importers for losses incurred (compared with a 'business-as-usual' scenario) as a result of CO_2 abatement measures to be adopted by their customers. The reaction to this demand in the West was negative, because of the potentially very large sums involved, the fact that the postulants are rich countries compared with most of the developing world, and above all the recollection that no compensation for importers had been on the table in the past when oil prices were very high.

In view of the uncertainties about how, and how vigorously, the Kyoto commitments will be applied by energy-importing countries, it is extremely difficult to forecast the impact of the Protocol on global oil and gas consumption. If oil and gas conservation policies are implemented in as slow and cautious a way as now seems likely, demand will probably still continue to grow for at least another 20 years, although at a slightly reduced rate compared with unbridled consumption. Although not the best of all worlds for the UAE, this would be completely manageable by keeping production capacity growth in line with demand.

There are, however, much bleaker scenarios for oil exporters, mostly constructed by extreme environmentalist advocates. In some of these, oil demand could be down to 70 per cent of current levels by 2010, and halved by 2020. This is very difficult to visualize today, but it is just possible to imagine extreme climatic and/or technological developments leading to drastic reductions in oil consumption within this period. Such a picture would be extremely serious for Abu Dhabi and other oil exporters.

One major purpose of Peter Kassler's RIIA study, as reported here, was to consider a possible agenda of options for discussion between oil producers and the rest of the international community, with a view to both rationalizing the energy management of the importers and sustaining the economic well-being of the exporters. These might include:

- an end to coal subsidies in industrial countries and the introduction of energy taxes genuinely based on carbon content of fuels, favouring gas and disfavouring coal, relative to oil;
- actions to increase gas exports from oil-exporting countries (many of which have large undeveloped gas resources);
- involvement of oil exporters in international greenhouse gas trading schemes, the 'currency' of which could include elements such as the reduction of emissions from oilfield installations themselves (an activity pioneered by the Norwegians) and the promotion of oil sales streams specifically backed up by environmental projects, such as forest plantations and selling at a premium price reflecting the CO_2 reduction achieved ('green oil');
- disposal of CO_2 in wells in Middle East oilfields, possibly linked to enhanced recovery of oil and natural gas;
- alternative energy developments in producing countries, including solar power generation schemes.

Developments of this kind will not happen tomorrow, nor need they, since the changes in the oil industry are slow and on a very long time horizon (although there may be surprises, as suggested above). However, it would be complacent to assume that oil will remain the predominant fuel for ever, and Kassler suggests that if change is inevitable it is never too soon to take the first steps to adapt to the new order.

There was some reflection of these thoughts in the discussion at the seminar. One discussant pointed out that consuming governments have great power to change the patterns of energy consumption, in favour of nuclear or renewables, for example, by means of taxes and incentives. Mention was also made of the growing influence of Green parties in European politics. Both these points suggested that change might be accelerated by non-economic factors. Consumers' preferences are changing. Some US consumers have shown themselves willing to pay extra for energy supplies from renewable or 'environmentally friendly' sources. This could be an opening for the 'green oil' idea mentioned above.

Another signal of potential change raised in discussion is the attitude of major coal-producing countries to warnings of climate change. Many commentators have been very pessimistic, feeling that they would go on burning their coal regardless of any pressures to the contrary. In fact, however, in the United States it is hard to visualize a major utility constructing new coal-fired capacity, because of the fear that coal was becoming 'the next nuclear'. In India and China, also big coal producers, there is considerable interest in increasing use of natural gas. For China this would largely be from Central Asia by pipeline, but both countries are also potential customers for liquefied natural gas imports. This could be a major opportunity for Gulf countries with surplus gas supplies.

The Emirates environment: ecology, water and the war against pollution

Environmental developments in the UAE

Saif Al-Ghais gave a detailed account of the environmental developments in the UAE that have taken place in the context of a spectacular growth of the economy and the oil industry over the last quarter-century, accompanied by an average annual increase of 10 per cent in the population, to a large degree by immigration, and by parallel growth of secondary and service industries throughout the Emirates. This huge expansion of the UAE economy has challenged the fragile desert and marine environments in many ways: by industrial and domestic pollution, by destruction of habitats through the growth of towns and cities, and by consumption of resources including water and fish stocks.

Al-Ghais' story is about the determination of the UAE, led by its president, H. H. Sheikh Zayed, to resist this destruction, and to make Abu Dhabi and the other Emirates greener and more friendly to plant and animal life than they were before oil came along.

In the Arabian peninsula, where people have lived at subsistence level for hundreds of years, utterly dependent on food and water wrested from hostile arid and semi-arid environments, traditional values do strongly encourage the conservation of natural resources. This message has been emphasized by Sheikh Zayed in his speeches and by the actions of his government, including the promotion of irrigated plantation, protection of terrestrial and marine wildlife, management of water resources and sewerage, and encouragement of environmental research and literacy.

In spite of the hostile hot, arid climate the national and municipal drive to 'green' the urban areas has been impressively successful. The attractive trees and bushes along roadsides improve the quality of life in the towns and cities as well as providing a habitat for native and migratory birds. The bulk of the water used for irrigation is treated sewage effluent. National wildlife programmes are aimed at the conservation or re-establishment of land and marine faunas which have suffered from the changes to their habitats that have occurred over several decades as a result of the growth of the oil and other industries and the housing and other needs of a much larger population. The ecology of the rich wildlife of terrestrial and marine mammals and reptiles, birds, fish and insects has been studied and documented (with many studies executed or led by Saif Al-Ghais). This work has led to the establishment of wildlife reserves on several offshore islands in which wildlife restoration programmes are being carried out.

Commercial fisheries are a particular area of concern. Here as in other parts of the world they have suffered from over-fishing and marine pollution, and a management system has been set up to make the fish stocks and their habitat more sustainable. This has included efforts to re-establish mangrove areas, which have become degraded and which are an important link in the marine ecological chain.

Another major element in the movement towards sustainability of the local environment in the UAE has been the leadership of Sheikh Zayed in establishing institutions such as the Federal Environment Agency and the Environmental Research and Wildlife Development Agency. The work of these agencies is important, not only because of the studies which they promote and execute but also because of their role in spreading public awareness of the need for conservation in everyday life. Saif Al-Ghais points out that the everyday use of renewable materials is rooted in the history of the Emirates, when people would use palm trunks and fronds to construct and roof their houses. The environmental agencies also take part in studying and monitoring the threats from pollution of land, air and water, and in setting up training and awareness programmes to gain public support for anti-pollution campaigns.

In summary, the UAE environment is fragile and has been threatened by rapid change, but Saif Al-Ghais has demonstrated the extremely robust response of the Emirates government and people to meet this challenge.

In discussion it was suggested that one might be pessimistic about the long-term outcomes of these conservation efforts, but Saif Al-Ghais recalled that desert people are accustomed to making difficult choices about how they use their water and other resources in as sustainable a manner as possible. The point, therefore, is to honour the traditional wisdom, since it is God who determines whether these human efforts will succeed or not.

Water shortages and conservation

The paper by **Roger Cragg and Hywel Thomas** reminds us that water shortage is a global phenomenon and one which is getting worse at an alarming rate. In the Gulf region there is so far a lack of long-term quantitative information on water resources and flows, and the authors make a strong argument for this to be improved. It is clear, however, that precipitation in the inhabited areas is very limited (approaching zero in parts of the desert), and that abstraction is increasing to supply the needs of growing populations, agriculture and industry. Aquifers are therefore being depleted and only partly replenished, the supply needs being made up by desalination of sea water and brackish ground water, and recycled effluent.

Only limited options are available to increase total water supplies. These include building more dams across wadis to encourage entrapment of rain water (although the effectiveness of this method will be reduced by evaporation), transportation of water from outside the region, and 'mining' of ground water from ancient, non-rechargeable aquifers which have not yet been developed as supply sources.

More desalination is always available in principle, and the Gulf states are lucky enough to have abundant energy supplies available for this purpose. However, the drawbacks include the very high cost.

In the discussion, reference was also made to the impact of increased fuel use on CO_2 emissions, in response to which Thomas suggested that renewable energy sources including solar and wind power could be used for desalination (in line with one of the sustainability options suggested in Peter Kassler's paper). There are also newer desalination technologies becoming available, including the 'sea-water greenhouse' based entirely on renewable resources.

Cragg and Thomas, however, also place great emphasis on the need for management of water demand. At the macro level this should include some difficult choices about the scale and types of agriculture and industry which can be sustainable from a water-supply viewpoint in the Gulf.

Agriculture in this region is extremely water-intensive, consuming about 70 per cent of supplies, and the products grown could in principle be replaced by imports from the global food market. However, like most sustainability questions this has more dimensions than just water supply; the sustainability of the communities producing the food also has to be taken into account, as well as macroeconomic consequences for the trade balance and issues of dependence on external supplies. Within existing patterns of food production there is probably scope to improve the efficiency of irrigation techniques to minimize evaporation. New techniques, including satellite remote sensing methods, are also available to monitor this type of water use.

One further possibility raised in discussion was the potential development through genetic engineering of salt-tolerant crops, which has apparently been considered in the region.

Industrial demand is much lower than agricultural, consuming about 5 per cent of supplies. Here, as elsewhere, there is a need for more measurement of water use to ensure its efficient deployment, and perhaps also a more searching application of cost–benefit analysis to the outputs of water-intensive industries in Gulf countries.

The remaining 25 per cent of Gulf water supplies goes to domestic use. Cragg and Thomas mention that the Gulf probably leads the world in the re-use of domestic water. Even so, there are a number of areas in which water management could possibly be improved or intensified. These include the control of leakage from the pipe system, the pricing of water, the installation of sea-water supply systems for non-drinking applications such as toilet flushing, and even greater re-use of domestic water.

Leakage management depends in the first instance on measurement of the flows in and out of supply systems and establishment of a standardized reporting system. There has been substantial improvement over the past few years in the technologies available to do this, and there may be a need for water suppliers, having access to the technology, to involve themselves in detecting leakage in and around consumers' domestic pipework. Free repairs may save more in water than they cost to undertake.

Water metering is widespread in the Gulf, but there are major social and economic issues around charging structures. There is a basic clash between the need for charges to reflect the value of a really scarce resource like water and consumers' need for the water itself, absence of which is life-threatening in a hot country. Cragg and Thomas outline some imaginative ideas about tariffs which provide stronger incentives for consumers to economize on the use of water for discretionary purposes such as filling swimming pools, without discouraging essential use.

There is considerable scope for micro-management of water use by the introduction of water-conserving domestic appliances – washing machines, dishwashers, cisterns and so on. Cragg and Thomas also quote the example of Hong Kong, where for many years there have been domestic supplies of sea water to high-density urban developments, mainly for flushing. In the Gulf this would, however, have a practical limitation because it would make the effluent less useful for the irrigation schemes into which it currently goes.

The underlying requirement for the introduction of water conservation and management practices such as those described is the presence of appropriate institutional and economic frameworks. Many of the ideas outlined are politically sensitive and would need strong leadership to put them in place. Underlying the whole question is a need for more measurement, reporting and assessment of water use. Cragg and Thomas leave us in no doubt that the potential water deficit in the Gulf is serious enough to justify these efforts. Ignoring industrial use, about three times as much as is currently used for agriculture would be needed for self-sufficiency.

One discussant questioned the role of privatization in introducing a more rational water policy. The response, quoting the UK example, was that privatization had led to a much stronger focus on the proper use of supply facilities and on the economic value of the product, in what had previously been a very old-fashioned industry. One issue whose profile had been greatly raised after privatization was the amount of wastage from the supply system. Coupled with public reluctance to support new reservoir construction for environmental reasons, this had led to real improvements in water management. This was not to say, however, that privatization was necessarily the best solution in all situations; it was one of a range of options available. Hywel Thomas also emphasized the central role of the regulator in assuring fair play between suppliers and consumers, particularly on pricing.

There was also discussion of international issues of water management. One very important issue for the UAE and its neighbours is the effect of very large water schemes, particularly in Turkey, on the volume of fresh water flowing into the head of the Gulf through the Tigris and Euphrates rivers, which will presumably greatly reduce the water circulation within the Gulf, with adverse ecological effects. No immediate solution was available to this potential problem, but it is clear that, as was the case with global climate change, transnational water sharing is a question on which public awareness is rising quite rapidly.

Managing the oil spill risk

Brent Pyburn's paper describes measures to minimize the ecological impact of oil spills in the Arabian Gulf. Underlying these is a dual logic: both to reduce the risk that spills will happen and, if they do, to deal with them effectively. A great deal can be done to raise the standards of ships and terminals, and their crews and operators, to the point where spills are extremely unlikely; but the ones that do occur in spite of the precautions taken are almost always the result of human error, which will always be with us.

On prevention, marine oil discharges may be either deliberate (disposal of contaminated ballast or washing water for commercial reasons) or accidental.

Deliberate discharge is a criminal offence under international conventions, which also stipulate operational procedures to prevent the disposal of oil or contaminated water offshore, including the requirement that ports provide shipmasters with contaminated-water reception facilities. The other side of this coin is the policing of marine traffic lanes by overflying them with aircraft fitted with oil monitoring equipment, which acts as a deterrent to 'rogue' shipmasters. This is currently done in the English Channel by the UK government.

In discussion, the funding of compensation programmes for damage resulting from oil spills was one question raised. This is covered by two international conventions providing for separate funds. One is the

International Oil Pollution Compensation Fund, financed by companies which import oil into a country. The other is the Civil Liabilities Convention, through which shipowners are required to pay a premium to insure their vessels and operations.

Clearly, however, although the legal and compensation arrangements exist, prevention is the only ethically responsible policy. Accidental discharges can be minimized by operating a marine assurance policy of vessel selection and auditing of terminals and operations. Brent Pyburn describes the policies and procedures used by BP Amoco as an example of how one very large and responsibly managed company does this. These policies are made up of the following elements:

- a strong policy statement emphasizing the company's commitment to 'no accidents, no harm to people and no damage to the environment';
- a policy to carry as much oil as possible in the vessels owned and managed by the company itself;
- a shipping audit policy covering the entire voyage but having a particular focus on its beginning and end, including loading and unloading, and transit of river and estuarine sectors under pilotage;
- a ship vetting policy applied to chartered vessels and their owners and managers, specifying operational practices and standards above international statutory requirements. Inspection reports on vessels are filed by members of the Oil Companies' International Marine Forum in an international database available to other members, to avoid repeated inspections of the same vessels. However, any evidence of a change in a vessel's ownership, flag of operation, or technical performance will lead to a review of its status.

One discussant questioned the degree to which oil-carrying ships at sea are able and prepared to deal with incidents. Here it appears that a great deal has been learnt from accidents which have occurred, and extensive procedures and arrangements are in place to deal with them. The industry therefore feels confident that it can operate its energy vessels effectively.

Notwithstanding all these procedures, accidents do occur. Brent Pyburn reports that recent major spillages were caused on relatively new and well-equipped vessels by highly trained professional mariners who did not follow strictly laid-down procedures. There is therefore a need for oil spill response processes to deal with such accidents. These include contingency plans, the provision of specialist equipment and resources, and a management commitment to complete the response in as short a time as possible with minimal damage to the environment. In BP Amoco this is based on an attitude of initial 'over-reaction' until the situation is fully understood, followed by reduction of the response to the appropriate level.

Contingency plans are needed to cope with a wide variety of foreseeable spill scenarios from small to very large. As for other types of plans, it is vitally important that they are kept up to date and that all those who may be faced with dealing with an incident are completely familiar with them.

A critical aspect of such planning is the interaction between the oil industry and the national authorities of areas where a spill may occur. Around international waterways there may be several countries potentially affected by any one such event. The national plan is therefore crucial in coordinating response to a major incident, as only governments have the necessary resources and authority to take the overseeing role, although the expertise may need to be drawn from the industry.

National preparedness is covered by the International Convention on Oil Pollution Preparedness and Cooperation (OPRC 90) adopted in 1990 by the International Maritime Organization. This provides for each signatory country to designate a responsible authority and contact point, to set up a national plan and international agreements with neighbouring states, and to coordinate responses with the oil and shipping industries and other parties which can provide technical advice and resources to help it to respond to an incident. OPRC 90 was followed by widespread industry and government initiatives to raise awareness of the need for contingency planning and to provide assistance to – particularly – developing countries in planning for oil spills.

An important part of the structured response to spills is the classification of occurrences into 'tiers'. Tier One incidents are small operational spills remediable by a local response and requiring locally available equipment. Tier Two incidents are more severe, either at a terminal but in excess of the capacity of local resources, or remote from facilities. These require some degree of combination of resources of different organizations and stockpiling of equipment at strategic localities.

Tier Three plans cover larger incidents requiring international cooperation. The Gulf Area Companies' Mutual Aid Organization is a good example, whereby, in the case of a catastrophic event in the Gulf, companies in the region will donate equipment and expertise. There are, around the world, a number of Tier Three centres, including in Florida, Singapore and the south of England, where equipment and dedicated staff are maintained in readiness to travel to a major spill and remedy it.

The nature of the responses to spillages has changed radically as experience has been gathered and digested from those which have occurred. In some cases, aggressive shoreline cleaning operations have done more damage to flora and fauna than they prevented. There have also been prejudices preventing the early use of chemical dispersants which could in some instances have dealt with the oil while it was still close to the vessel; in the event it spread much more widely than was necessary. With very light oils and turbulent seas, nature can provide the best response to a spill; the example cited is that of the *Braer* off the Shetland Islands in 1993, when rough seas very rapidly dispersed the oil into the water column for biodegradation.

More recently, the technique of net environmental benefit analysis (NEBA) has been used with the aid of environmental specialists to determine a minimum damage strategy for each spillage as it occurs, in the light of the specific circumstances and the types of animal and plant life at risk. Brent Pyburn gives an account of the technologies and management systems available to deal with pollution incidents. The challenge remains, however, to be able to establish a rapid consensus about the correct course of action among the various parties involved (governments, local populations, companies) under conditions of uncertainty and, often, extreme political and media exposure.

Conclusion

The Abu Dhabi seminar in October 1998 set out 'to examine the implications for the United Arab Emirates of global environmental concerns and measures to address them'. This was an ambitious objective, reflected in the broad spread of the topics covered, extending from oil to water to nature conservation, with approaches ranging from the global to the very local.

The individual issues are not new – debates about the management of petroleum and water resources and the fragile natural environments of the Gulf have been going on for decades and will not quickly be resolved. What is new in the work reported in this book is an attempt to understand and promote sustainable development as an integrating idea existing at a higher level than the familiar day-to-day problems facing governments and corporate managements. This implies a new search for solutions to such problems which will satisfy sustainable development needs as well as the sectoral ones of immediate concern to civil servants and corporate executives (Roger Rainbow's 'triple bottom line'). Because sustainable development issues so clearly cut across the boundaries of both economic sectors and nation-states, it also implies a search for compromises and consensus among a wider range of actors than is currently the custom. In this context the notion of a hierarchy of sustainable development agendas – global, regional and local – should be helpful in creating a structure of problem-solving and decision-making within which individual economic and environmental topics can be located.